The MIT Press
Cambridge, Massachusetts
London, England

Lamarck the Mythical Precursor

A Study of the Relations between Science and Ideology

Madeleine Barthélemy-Madaule

English translation by M. H. Shank

This book was set in Sabon by Achorn Graphic Services, Inc. and printed and bound by Murray Printing Co. in the United States of America.

Library of Congress Cataloging in Publication Data

Barthélemy-Madaule, Madeleine.
Lamarck the mythical precursor.

Translation of: Lamarck ou le mythe du précurseur.
Bibliography: p.
Includes index.
1. Lamarck, Jean Baptiste Pierre Antoine de Monet de, 1744–1829.
2. Evolution. 3. Naturalists—France—Biography. I. Title.
QH31.L2B3613 1983 575.01'66
82-10061
ISBN 0-262-02179-X

Contents

Foreword

It is not Lamarck and neo-Lamarckism, but rather Darwin and Darwinism that make the headlines these days. In the last few years, journals like *La Recherche* have devoted issues to Darwinism (e.g., Jören Lövtrup's 1977 article on the "Crisis of Darwinism"). Yet many scientists, some of whom will be quoted in this book, are questioning it. The heredity of acquired characters is becoming what Pierre Viansson-Ponté dares to call "the great debate."[1] The environment claims its rights on organic modifications, even though the genes do not surrender. Behind these questions stands the enigmatic—and growing—shadow of Jean-Baptiste de Lamarck. Why does Lamarck occupy such a prominent place in recent books by well-known scholars such as Yvette Conry and Camille Limoges? Why did H. Graham Cannon recently devote a book to *Lamarck and Modern Genetics* (1959)? Why has Richard W. Burkhardt published a monograph in 1977 on "Lamarck and evolutionary biology" entitled *The Spirit of System*? Lamarck has evidently become a point of reference for an international movement. When we raised questions about the relation between science and ideology with a group of graduate students in the *troisième cycle* from the University of Picardie, we inevitably encountered Lamarck. So it happened that although I was involved in patently historical research, I suddenly found myself in the midst of a topic of current interest.

I wish to thank my colleague in this *troisième cycle* course, Dominique Lecourt, whose provocative questions and precise comments proved most helpful. I also wish to thank Albert Jacquard, professor in Paris VI and research director at the Institut national d'études démographiques, who generously referred me to valuable documents written by himself and by his associates in Paris and Geneva. Fathers Dumortier and Brunet, from the splendid library of the Centre culturel des Fontaines, put relatively inaccessible books at my disposal. I thank them, as well as Father Russo, who was so kind as to grant me access to documents in his possession that deal with several very recent issues on the question of Darwinism and acquired characters. I shall restrict myself to these immediate gestures—to add to them, it would be necessary to mention too many names. Nevertheless, they are very much present in this study. Yet I cannot forget to thank Yves Laissus, director of the library of the Museum d'histoire naturelle; Philippe Dupont, director of the university library in Amiens; and G. Tournouer, director of the municipal library in Amiens.

Finally, to avoid any misunderstanding, I would like to make my purpose clear. I set out not to write an exhaustive account of Lamarck's life and work, but rather to treat the reciprocal relations of several key Lamarckian concepts in order to shed light on the relation between ideology and science.

Preface: The Myth of the Precursor

Lamarck's destiny is strange indeed! He climbed the rungs of a seemingly brilliant career for which his colleague and adversary Georges Cuvier did not fail to reproach him in a eulogy that remains a model of the vicious academic attack. He died in misery and oblivion. Even the naturalist Etienne Geoffroy Saint-Hilaire, who shared Lamarck's transformist convictions, dared to praise him in his funeral oration only for work unrelated to the *Philosophie zoologique.* He was forgotten in spite of the place that men like H. D. de Blainville, Auguste Comte, Arthur Schopenhauer, and even Honoré de Balzac granted him. He was eclipsed by Darwin. Alfred Giard, however, thought that Lamarck was reborn from his ashes whenever the ideas of progress and liberty held the upper hand, in 1830 and 1848, for example. Still numb from its recent defeat, the France of the 1870s found solace by seeing in him the pioneer of evolutionism. Yet Ernst Haeckel's Darwinism did not prevent him from doing justice to Lamarck, to whom would "always belong the immortal glory of having for the first time worked out the Theory of Descent"[1] at the beginning of the nineteenth century. At the beginning of the twentieth, a certain free-thinking current represented by Alfred Giard, Félix Le Dantec, and Edmond Perrier, among others, thought it saw its reflection in Lamarck. The alternating patches of light and darkness that characterize his life and reputation were attributed to his incomparable genius far ahead of its time and the ill will of the fearsome Cuvier, the defender of transformism: "It sometimes requires a high price to be a man of genius—and the farther ahead of his times he is, the higher the price";[2] the conspiracy of obscurantist forces had done the rest. It is in these terms that the current favorable to Lamarck constantly promoted him to the incomparable rank of precursor.

But what specifically is a precursor? In 1966 the historian of science Georges Canguilhem alerted his readers to the ambiguities of the term.[3] His comment about the expression "the past of a science"[4] also seems to fit the expression "precursor," namely, it is a "vulgar" concept. Moreover, "precursor" and "time" are correlates. "A precursor is presumably he who runs ahead of all of his contemporaries, but also he who stops on the track where others after him run to the finish line."[5] Thus the precursor in his own time lives a time that does not yet exist. Although he seems to dominate the temporal dimension, in fact he belongs to no instant. He has been detached from the structures and achievements of the past of his period thanks to, and by the fault of, an almost prophetic vision. But since he has come too soon, "before his time," he does not belong to the period in which it is held he fits, for the structures of its knowledge have no direct connection with his thought. Caught between "not yet" and "no longer," where is the precursor's time? Uchronia, utopia! His existence raises questions about the "historical dimension" of a science.[6] Understood in this sense, Canguilhem has harsh words for the notion of precursor, which is so difficult to conceptualize as anything but an illusion. But even if one shares this evaluation, should one not raise the question, Why have so many well-known scholars and scientists used the term, and specifically for Lamarck? The majesty of its content obscures its imprecise status. But it also perpetrates an injustice on the scientist one intends to glorify, for his greatness is then based on an "elsewhere" that uproots him from his context and from the work he produced in it, and is therefore unsuited to throw light on what characterizes him.

François Jacob claims that the history of biology has two modes. The first, which focuses on the "succession" and the "genealogy of ideas,"[7] makes us fall into what he terms "reverse history," produced by "the extrapolation of the present toward the past." Canguilhem, who also noticed this form of retrospection, asks in this connection, "But is this science of the past the past of today's science?"[8] Are not the so-called antecedents uncovered as a function of "the needs of the day?"[9] It is only in the perspective opened up by this question that "Lamarck can be seen as a predecessor of Darwin, Buffon of Lamarck, Benoît de Maillet of Buffon, and so on. But this fails to explain why Lamarck's ideas were almost totally ignored at the beginning of the nineteenth century by men such as Goethe, Erasmus Darwin, or Geoffroy Saint-Hilaire, the very people who were seeking arguments in favor of transformism."[10] The problem is very aptly formulated here: On the one hand, it unveils the arbitrariness of the past, and therefore of history itself; on the other, it warns the historian of science about the recurrent genitive that always makes person A (in the nominative) the support of person B (in the genitive) by giving A an identity other than his own (namely, that of B), an endless "to be continued," which robs the precursor of his own substance.

As for Jacob's second aspect of the history of biology, it is reminiscent of Michel Foucault's *Archeology of Knowledge*, according to which every period unfolds its field of possibilities and their synchronic relations. Thus every producer and every product find their own specific density. The precursor loses his fictitious function and remains only a producer of what he effectively accomplished, and would have been in any case. As Jacob says in substance, if Darwin had not been there, Wallace would have taken charge of evolutionism as a truth and not as the fictionalized account of a discovery. In this concept of a domain to be explored, Lamarck is truly Lamarck. It is our burden to ask, What is the balance of his effective production? His itinerary and his procedure are no longer put into continuity with those of Darwin at the price of erasing their differences. What concepts did Lamarck have at his disposal? What new concepts did he elaborate? To what notions did he appeal to fill in the blanks? In the introduction to his *Études d'histoire et de philosophie des sciences*, G. Canguilhem quotes Alexandre Koyré as follows: "It is nevertheless evident—or at least it ought to be—that no one ever thought of himself as someone else's precursor, and indeed could not have. Hence to consider him as such is the best way to fail to understand him."[11] Now in order to determine Lamarck's scientific status, we must denounce the "vulgar" level of notions that gives rise to myths like the notion of precursor, which has no logical rigor but only a type of narrative temporality of an imaginary order not perceived as such.

The myth of the precursor springs from the same source as the myths of political history. The point is not to exclude them, but rather to assign to them their proper logical and chronological places. At the end of the nineteenth and at the beginning of the twentieth centuries some of the great thinkers, philosophers, and scientists spread the myth of the precursor. In his introduction to the *Philosophie zoologique*, Charles Martins emphasized the visionary perceptiveness of Lamarck's genius. In his introductory lecture for 1888, Giard sang Lamarck's praises and bemoaned the fact that "in this nineteenth century, in which all of the sciences grew so rapidly, ideas such as these took eighty years to cover the distance that separates the Museum [d'histoire naturelle] from the Sorbonne, and they did so only after a circuitous voyage through England, Germany, Russia, and even America."[12] And a few lines later, he added, "Like Lamarck, Darwin also had his precursors."[13] In his *History of Creation*

(1868), Haeckel wrote, "How far it was in advance of its time is perhaps best seen from the circumstance that it was not understood by most men, and for fifty years was not spoken of at all."[14]

All the elements of the myth are in place: the misfortune of the unrecognized genius; the ingratitude, stupidity, and ignorance of men; the maleficent power of the adversary Cuvier; and fate, which envelops the whole. The story line fills the explanatory vacuum: The great figure of a Lamarck pursued by fate finds expression in the statue near the Museum, where he worked so much! Lamarck "sad, unrecognized and despised,"[15] pursued to death by fate! "Cuvier, the enemy of the new biology, was at the peak of his glory and honor when Lamarck, poor and forgotten, was buried in a plot of the Montparnasse cemetery, from which his bones would soon be removed because he could not afford a perpetual tomb!"[16] Edmond Perrier, who before his death in 1925 picked up on the captivating myth with emotional solicitude, raised the important question, "Why did he remain in the shadows?" His book on Lamarck presents itself as an answer under the guise of a story that "deserved to be told"[17] and that he too found surprising: "It is astonishing that with such a creative genius, Lamarck remained almost completely unknown in his own time."

Lacaze-Duthiers, the successor to the chair vacated by Lamarck and de Blainville, noted in a letter to A. Packard that Lamarck's colleagues called him "a fool."[18] To be sure, there were attempts to answer the question, "Why the silence, why the oblivion?" Cuvier's ill will did not explain all, as Lamarck's defenders knew very well. They too unmasked the ideological battle, the hostility of the creationists, and of the reactionary forces. And Lamarck appears as the enemy of religion and teleology, even as a mechanistic materialist. But the same apologists failed to point out that the theological offensive against Darwin engendered neither silence nor oblivion.

Lamarck's peculiar fate raises a question at the most existential level. And to this question probably corresponds a theoretical problem. Has Lamarck been sought where he really is, awaiting the recognition that comes from cognition? The aim of this book is to contribute not an answer, but only a few of the elements required for a clearer understanding of the "Lamarck problem."

Lamarck the Mythical Precursor

1 Who Is Lamarck?

His life, his ideas, his personality—why should they matter to us? Where do the anecdotes end and the outlines of a meaningful structure begin to emerge? It may be interesting to consider his astonishing destiny. An aristocrat born into a family with few options was granted honors under the Convention. In his devotion to equality and his enthusiasm for the times in which he lived, in the dedication of his thoughts and discoveries "to the French people," he was no opportunist trying to save his life and his career by flattering the masters of the day. Later, under the Restoration, his *Système analytique des connaissances positives de l'homme* (1820) would display the same tendencies and express the same ideas. At the turn of the nineteenth century, Lamarck found himself in the midst of many a transformation, perhaps of many a contradiction. The man who followed the aging Jean-Jacques Rousseau on his field trips was also Darwin's predecessor. He mediated between the Era of Progress and the Industrial Age, and stands as a witness to both their contrast and their solidarity.

The risk of remaining at the anecdotal level is small in discussing a man who, according to his biographer Alphaeus Packard, left neither tomb, nor letters, nor manuscripts.[1] In the late nineteenth century, Packard traveled from Brown University to France with the explicit purpose of finding every possible trace of the man. He went to Bazentin-le-Petit, Lamarck's birthplace in the *arrondissement* of Péronne, canton of Albert, in Picardy. Thanks to a local school teacher, Duval, he was allowed to visit the large building,[2] pompously called a castle in spite of its houselike style, where on 1 August 1744 Jean-Baptiste Pierre-Antoine de Lamarck was born to *messire* Jacques de Monet, *chevalier* de Lamarck, lord of Bazentin *grand et petit*, and to "high and powerful lady" Marie Françoise Fontaine. So reads the birth certificate recorded at the town hall.

Nothing is known about the little boy who would later become the founder of transformism, except perhaps that "from childhood, he enjoyed solitude and showed little interest in the inclinations of his peers," as de Blainville and Maupied allege. That he grew up during the Enlightenment is far more interesting. In 1744 Rousseau had not yet published any of his great works. Lamarck was five years old when the discourse entitled "Whether the reestablishment of the sciences and the arts has contributed to the purification of mores" came off the press. He was ten when the Amsterdam publisher Marc Michel Rey received the manuscript of the *Discours sur l'inégalité*. That same year, Condillac published his *Traité des sensations* and Diderot his *Considérations sur l'interprétation de la nature*. During this time, Charles Bonnet was studying nature and proposed in writing the "chain of beings." Many more great names and works in science, philosophy, and literature mark the years of Lamarck's childhood and youth. When Rousseau's *Emile* and *Du contrat social* came off the press, he was eighteen.

Meanwhile, Pierre de Maupertuis, Jean d'Alembert, the *abbé* Nollet, and others were devoting themselves to physics and astronomy. Newton's name was surrounded by glory. In a field that would soon be called "biology," Trembley was studying the fresh-water polyp while Buffon was beginning his *Histoire naturelle*. Thus, although the bulk of Lamarck's work belongs to the early nineteenth century, he was through and through a man of the eighteenth century, the man of the Encyclopedists, but also of the Ideologues.

But after considering the narrow circle of his family, one might guess that he traveled a long road before realizing the significance of the Enlightenment. What indeed could have been the mind set of the family that produced Lamarck? On his father's side, he seems to stem from the old nobility (middle nobility according to some; high nobility according to others). Although originally from the Béarn, they had settled in Picardy long enough for the name La Marque to have changed into Lamarck, with its Flemish ending.[3] Several sources shed light on these lords of Bigorre: a family tree constructed in 1757 (although destroyed during the French Revolution, one copy survived in the papers of a relative);[4] the papers collected by Cuvier for the eulogy he was to deliver before the Académie des sciences;[5] several periodicals;[6] and finally the archives of the Museum d'histoire naturelle.[7] Lamarck's forebears had been port commanders and governor-officers, worthy ancestors of his three older brothers: the eldest was killed during the siege of Berg-op-Zoom (1747); the other two served in the army as well. The century of Louis XV supplied enough wars to keep the family busy. As we shall see, Lamarck was also faithful to this tradition. On his mother's side, the Picard line can be traced back to the tenth century. The Fontaines were descendants of the Maieurs d'Abbeville, counts of Ponthieu, a lineage that stretches back to the Crusades.

But Lamarck's father had decided that his eleventh son would enter a religious order. The penniless nobleman was therefore sent off to the Jesuit school in Amiens, a far cry from the intellectual aristocracy imbued with bold new ideas in philosophy, science, and politics. Between the military atmosphere of his home and the climate of his school, Lamarck was in no danger of being swept away by the winds of the century. His training included dead languages, logic, scholastic philosophy, and mathematics.

(It *can* take credit for his physicochemistry.) The archives of the school disappeared during the Revolution. We know only that the populace of Amiens nicknamed the pupils *capettes* on account of the collars they wore. Lamarck did not enjoy being there and seized the first opportunity to leave. Almost all of Lamarck's biographers link his departure with his father's death. But the latter died in 1759, as Marcel Landrieu has pointed out, and the son returned home only one or two years later.[8]

Lamarck's biographers have all dwelt on the military episode of Willinghausen in Westphalia. Their common source is a letter addressed to Cuvier by Auguste de Lamarck shortly after his father's death—a document with an obvious academic purpose. What are we to make of this event? Is it a youthful character trait, a family trademark, a manifestation of qualities that will later be put to the service of science and of the difficult struggles that make Lamarck's existence continuously heroic? In July 1761 Lamarck is sixteen years old and living in Bazentin. His mother finally gives in to his pleas and grants him permission to enter a military career. Riding a cheap horse, with a turkey-keeper for squire and a letter of recommendation in his pocket, the aspiring d'Artagnan sets off on the long road toward the armies of the Seven Years' War. The newcomer does not exactly enchant the colonel de Lastic. Nevertheless, taking advantage of the latter's summons to headquarters, Jean-Baptiste de Monet de Lamarck takes charge of a company of grenadiers. When the commander returns and orders him "to follow the supplies," Lamarck replies that he has come "to serve" and that the braves "will have no reason to be ashamed" of him. After several hours under fire, the company and its commanders have been decimated. Fourteen men remain. The army retreats. The small troupe is forgotten. No orders are sent out. The boy takes charge and refuses to move. Much later, the colonel discovers the group during an about-face of the French army. Lamarck is promoted to the rank of officer that very evening on the battlefield. His military career is off to a brilliant start. From 1762 to 1768, he lives with the garrison in Monaco, occupying his time with music and botanical field trips, as is the fashion in the France of Jean-Jacques Rousseau. One jolly day, a comrade picks him up by the head. He does not recover until Tenon, a surgeon of some fame, performs a serious operation. Lamarck's military career has come to an abrupt end. Another life begins.

Lamarck moves to Paris to receive the necessary medical attention. He brings with him a decided interest in botany and Chomel's *Traité des plantes*, which he has received from his brother in exchange for a music score. He decides to study medicine and botany. To supplement his modest living allowance, he works for a bank. He attends the lectures at the *Jardin du Roi*. He visits his brothers, who live in a country house near Mount Valérien. In short, hounded by poverty, he leads a rather aimless life until he meets Bernard de Jussieu, a great figure who polarizes the young man's uncertain tendencies. From that time on, Lamarck's scientific career begins to take on a clear shape, and the first stage is botany. Meanwhile, from the window of his room on the Montagne Sainte-Geneviève, he observes and classifies the clouds with the hope of discovering some unchanging causality for forecasting the weather. These meteorological interests continue unabated throughout his later work in botany, zoology, and biology; from 1799 to 1810 he would publish the *Annuaire météorologique*. He eventually abandoned publication, but only after a deplorable episode in which the brutality of Emperor Napoleon I humiliated the old scientist in a public scene from which the latter never recovered. Napoleon did not like Lamarck. He preferred men whose thoughts threatened neither the Bible nor the "ruling classes."[9] Baron Cuvier did not miss the opportunity of poking fun at his colleague's meteorological researches.

However that may be, the period from 1768 to 1778 is marked by the association with Jean-Jacques Rousseau, the influence of Jussieu, the protection of Buffon. Lamarck's *Flore française* was published in 1778, the result, as rumor has it, of an argument and possibly of a bet. The fact that he completed the work in six months does not enhance its seriousness. But one should also take into account the slow maturation of his botanical interests and competencies, which were put to use during the eighteen long years of work with Bernard de Jussieu. At this point, fantasy gave way to the conjunction of research into the natural diversity of plants with a desire to offer the layman a convenient means of identifying every plant. Linnaeus's artificial systematics was being superseded, without its gains being rejected, however. Here too Cuvier tried to play down Lamarck's accomplishments by alluding to the "latest fashion of science." The undertaking was of a different order.

As early as 1778, the problem of species, closely linked to the problem of evolution, came to the fore at the botanical level. After having made amends in a response to the reprimand of the Faculty of Theology (1753), Buffon had thought it necessary to proclaim solemnly that "animal species are separated by intervals that Nature could not cross."[10] But much later, in his *Discours sur la dégénération des animaux*, he asserted that the two hundred species he had catalogued could indeed have emerged from a small number of principal familes. In like fashion, Lamarck reached the notions of variability and transformation in his analysis of plant species in the *Flore*, and later of animal species in the *Système des animaux sans vertèbres* (1801) and in several of his opening lectures (1803 and 1806 in particular). "One is not truly a botanist because one can identify at a glance a large number of diverse plants, even according to the latest nomenclatures," he wrote in 1806, thereby relativizing the value of classifications and superseding *the* classification of Linnaeus. In his funeral oration, Etienne Geoffroy Saint-Hilaire attributed the origin of the *Flore française* to a synthesis of Jussieu's teachings and the Linnaean sexual system with the method of Tournefort. In his articles for the *Dictionnaire de botanique*, Lamarck would resolutely side with Jussieu.[11]

Daubenton wrote the preface to the *Flore française*. Buffon protected the author who, thanks to the former's efforts, published his work at the *Imprimerie royale* (king's printing shop) and was elected to the Academy of Sciences in 1779. But Buffon would not protect just anyone, and this should be kept in mind when evaluating Cuvier's insinuations in his eulogy.

These first efforts, this protector, these teachers all presuppose an illustrious institution only mentioned so far in passing: the *Jardin du Roi*, today the *Jardin des Plantes*, or Botanical Garden, surrounded by streets and squares named after, among others, Cuvier, Buffon, Geoffroy Sainte-Hilaire, Tournefort, Lacépède, and Jussieu.[12] The *Jardin du Roi* was conceived and realized in part during the reign of Louis XIII by Guy de la Brosse on the model of the *Jardin des Simples*, or garden of medicinal plants, in the faubourg Saint-Victor. The great-nephew of Guy de la Brosse was Crescent Fagon, praised by Saint-Simon and Fontenelle, a curator and lecturer at the *Jardin du Roi*, a great botanist, a good chemist, skilled in medicine and surgery, a protector of Tournefort, who was responsible for bringing together the imcomparable shell collection so dear to Lamarck. The *Jardin du Roi* prospered. Buffon became its superintendant and expanded it to the banks of the Seine. From 1739 to 1788 he spent several months a year on location and, according to Perrier, deserved to be called its "new legislator and second founder."[13]

Lamarck was now famous. Thanks to Buffon, he obtained the position of "royal botanist, with the responsibility of visiting foreign gardens and botanical museums and establishing close contacts between them and the *Jardin des Plantes*." In 1781–1782 he visited several European countries (including Holland, Hungary, Austria, Hanover, and Saxony) in the less-than-pleasant company of Buffon's son. Many years later, the elderly Lamarck could not recall without a violent outburst of emotion the occasion when the young M. de Buffon, who wished to go out alone one night, poured ink over his traveling companion's clothes. For in the eyes of this young man, the son of the Count of Buffon, Lamarck carried no weight—he was a gentleman, to be sure, but from a slightly proletarianized family. Humiliated, without protection (which amounts to the same thing), he found himself in a situation that convinced him to revolt against human inequality. From the outset, he held the modest post of keeper of the herbarium. In 1790, after the superintendent la Billarderie removed him from this position, he referred to himself as the "king's botanist, associated with the *cabinet d'histoire naturelle.*" That same year, he lost Marianne Rosalie de la Porte, his wife since 1782 and the mother of their six children. It was probably in 1790 that he published the *Mémoire sur les cabinets d'histoire naturelle*, his first reform proposal. Regnault de Saint Jean d'Angély defended the petitioners before the National Assembly. In 1791, Bernardin de Saint Pierre became superintendent of the *Jardin des Plantes.* He did not get along with Lamarck, who was put in charge of insects and worms and turned more and more to zoology.

Thus begins a story that ends in June 1793 with the creation of the Museum d'histoire naturelle. At Lamarck's suggestion, the Convention had studied the proposal to reorganize the *Jardin des Plantes*. Daubenton had given Lakanal the 1790 memorandum prepared by Lamarck, and the project was adopted. This episode raises a twofold problem: the cultural accomplishments of the Convention and Lamarck's relations to the French Revolution.

The historian Louis Blanc has called these accomplishments "work amidst the din of war," which he describes as sowing with one hand while striking with the other.[14] Although 1793 was a very turbulent year, the Convention was preparing the "establishment of the Ecole polytechnique and the Ecole normale," promoting the arts and the sciences, standardizing the French language throughout the country, working toward the establishment of the telegraph, ordering a draft of the Civil Code, inventing the decimal system and standard weights and measures, and reforming the calendar. On 26 June 1793 Lakanal presented to the Convention a proposal for national education. Robespierre was delivering his speech on education at the very moment Charlotte Corday was assassinating Marat. "Amazing synchronisms!" exclaims Louis Blanc: "On July 27, at the very height of the emotionalism generated by the crime and the execution of Charlotte Corday, the Convention orders the opening of the Museum, appropriates a yearly budget for the purchase of paintings and sculptures from private sales, and houses the arts in the palaces of kings."[15]

The trial of the Girondins and the Lyon insurrection coincide with steps taken to organize society. These are the days of the counterrevolutionary assaults from within as well as without, the days of the Committee on Public Safety. Meanwhile, at the Museum, Bernardin de Saint Pierre is fired and Daubenton named director. Twelve chairs are created. Lamarck, who is in charge of the library, is appointed to the chair of zoology. Etienne Geoffroy Saint-Hilaire is assigned to the vertebrates, René Desfontaines and Antoine Laurent de Jussieu to botany, Daubenton to anatomy, and so forth. Lamarck eventually turns his attention to "white-blooded animals," that is, to "invertebrate animals"; he officially abandons botany. This already memorable year is also marked by Lamarck's second marriage and the birth of his seventh child.

Lamarck's attitude toward the French Revolution is difficult to determine. The Lamarckians of the late nineteenth and early twentieth centuries thought of him as a republican with progressive ideas. In the preface to his edition of Lamarck's *Discours d'ouverture*, Alfred Giard remarked that "our great zoölogist" came back into the limelight whenever "an era of liberty" dawned. Alphaeus Packard thought that Paris had never been so merry as during the Terror, and that 1793 was a great time for enthusiastic young men. Nothing could have been more natural than for Lamarck to express his gratitude to the French Revolution, which had given a vigorous impulse to the Museum. Can one penetrate further into his political ideas? Consider the dedication of his *Recherches sur les causes des principaux faits physiques* (1794):

Accept, o magnanimous people, victorious over all of your enemies, people who has been able to recover the sacred and indefeasible rights of nature; accept, I say, not the fauning homage that groveling slaves addressed to kings, to ministers, or to the powerful who protected them, but the tribute of admiration that your virtues and energy, developed by the wisdom and intrepid constancy of your representatives, have earned you. Accept also a work, the fruit of many meditations and researches that can, from a variety of perspectives, become very useful to all of mankind, that can lead to the most valuable discoveries, be they in the ordinary labor of life or especially in the principal aspects of the art of healing, a work that I wrote only with these purposes in mind, and with which I do homage to you, both out of affection and from my desire to share glory, at the very least, according to my feeble powers, by making myself useful to my fellowmen, my brothers, my equals.

Is this a piece suited to the circumstances? If so, its tone is somewhat exaggerated. But there is no good reason to doubt the sincerity of this ardent period. Moreover, Napoleon's disfavor and the neglect Lamarck suffered during the Restoration make it sufficiently clear that he was not in the good graces of the post-revolutionary authorities. Monsieur de Monet de Lamarck, now the citizen Lamarck, was to remain a man of the Revolution. The deism of his scientific writings, the allusions to the "Supreme Being" or the "Supreme Author" are not enough to make him a fervent Catholic. The young man who had been pushed toward an ecclesiastical career and who later became the *protégé* of the Count of Buffon may have inherited the secret irreligion of his mentor, who told Hérault de Séchelles at Montbard in 1795, "The people need a religion. I always mentioned the Creator in my books, but one need only omit the word and replace it with 'the power of nature.' . . . When the Sorbonne quibbled with me, I was accommodating and gave them all the assurances they wanted. It is only banter, but people are dumb enough to be content with it."[16] *Creator sive natura*! Buffon's covert atheism probably went beyond Lamarck's own position, which seems closer to the deism of Rousseau—whether or not by way of Robespierre—than to the views of some Encyclopedists.

But the decisive argument in favor of Lamarck's profoundly republican ideas is the *Système analytique des connaissances positives de l'homme* published in 1820, during the reign of Louis XVIII. Although it was not a conformist work, there can be no doubt that he asserted the existence and omnipotence of God. On what basis would this conclusion be justified? The careful distinction he draws between creation and production is precisely the distinction between God and nature.

One should not forget, however, that this God has neither moral attributes nor any purposes known to man. He is the limit of thought, the first principle. He can be found in all of Lamarck's works, but there are no connections between Him and man. The moral universe is dominated by politics and jurisprudence—it is the universe of Rousseau's *Contrat social*, governed by the general interest and public happiness. And this happiness can only be the "welfare of the totality of the members of society"[17] viewed from the perspective of their equality. This happiness grows out of Rousseau's dream and Saint-Just's struggle. Lamarck inscribes the psychological within a political framework. His hierarchy of inclinations depends essentially on social position (oppressor or oppressed). The antidote to oppression is true knowledge. False knowledge leads to abuse of power. To spread knowledge, to know nature at every level, from the cosmos to man, must make for human progress. This is indeed the ideal of the Enlightenment. The entire *Système analytique* grows out of it. Lamarck is the man of the French Revolution as the fulfillment of the ideas of the eighteenth century. He is not the man of kings and emperors. The French Revolution granted him recognition. The emperor and kings did not. Perhaps they understood him too well.

After 1793 Lamarck's biography merges completely with the production of his works, right up to 1819, when after completely losing his sight, he probably dictated the *Système analytique* to one of his daughters.[18] Meanwhile, he married for the third time. His last marriage was childless. According to Edmond Perrier, three of Lamarck's eight children died in infancy.

In 1795 he became the secretary, and three years later the director of the Assembly of Professors. Apart from two or three short absences, he scarcely left the Museum. In 1809 he refused the chair of zoology in the Faculty of Sciences of the Sorbonne in spite of the financial advantages such a position implied. The devotion of his daughters is well known; one of them stayed with him until his death and worked on after him, labeling samples in the Museum for which her father had done so much.

Whereas Lamarck's birthdate is settled by town-hall records, his biographers cannot agree on the date of his death. According to Landrieu, he died on 18 December 1829 and was buried on the 20th. A Dr. Mondrière allegedly found the death certificate. Should Packard's date of 28 December therefore be rejected, all the more so because it coincides with a typographical error? (The burial date he gives is 23 December.) Yet Edmond Perrier has Lamarck dying on 25 December 1829, and Mantoy chooses 26 December! Packard tried to find the place where Lamarck was buried following the ceremony at Saint-Médard. I say "place" instead of "tomb" for Lamarck was so poor that he could not afford a plot. The authorities did nothing to alleviate the situation, in spite of the high position of Baron Cuvier, and the body turned to dust in some common grave. The devoted Packard nevertheless wanted to find the very spot in the Montparnasse cemetery where the coffin was put into the ground, not far from a certain Dassas in the first division of plots, a few feet from the large walkway. Perhaps his shadow ought to be sought instead in those "natural history cabinets" where he worked well beyond the call of duty, in the house owned by Buffon where he lived and died, and in those walkways of the *Jardin des Plantes* where he would meet colleagues who thought him slightly mad. "But then," as one of them said, "it is quite difficult to know precisely where he lived—he took up so little room among us."

Geoffroy Saint-Hilaire, the most illustrious of those who spoke at the graveside, deemed Lamarck "one of the greatest zoological talents of our age" and called him "the French Linnaeus." But this title was inadequate—it described the classifier of invertebrates, not the systematic thinker. Geoffroy Saint-Hilaire was more explicit after Cuvier's death. The latter had prepared the academic eulogy, that "thrashing" from which colleagues had begged him to delete several lines. Cuvier had refused, so that following his death it was another who on 26 November 1832 read his speech with its references to "fanciful notions," to "vast edifices" built on "imaginary foundations . . . not unlike those enchanted palaces that, in our old novels, can be made to vanish by breaking the charm on which their existence depends." History has reversed this interpretation: Lamarck is more famous for his *Philosophie zoologique* than for his classifications and observations; and Cuvier is remembered for his peerless analyses in comparative anatomy, not for the ideology or the rather "unspontaneous" philosophy[19] that prevented him from accepting evolutionary theory. It must be admitted, however, that Cuvier probably would have bowed down before the evidence.

Lamarck's failures in the eyes of many of his colleagues[20] and Darwin's haughty refusal to recognize him present a problem. A mixture of glory and disdain surrounds his name. Even during his lifetime he was a controversial figure. After his death, alternating patches of light and darkness, of glory and rejection highlight the "Lamarck problem." This is the question that this book will attempt to make more precise by rejecting the mythical answer of the precursor. Let us examine his work and the image it projects throughout the series of his publications. Let us take a broad view of this work, a view intentionally omitted from the preceding brief biography. Finally, after deliberately overlooking secondary considerations, let us concentrate on the heart of the matter, on the focus of the debate—transformism, which is its foundation—and ask ourselves what Lamarck accomplished when his works are seen in their historical context.

If we overlook the unpublished memoir of 1776,[21] where Lamarck vented his meteorological preoccupations, we see the great botanical studies developing. The *Flore française*, a synopsis of plants, the *Dictionnaire de botanique*, various memoirs, the *Journal d'histoire naturelle* succeed each other in the period from 1778 to 1790. It is surprising that shortly after his accession to the chair of zoology at the Museum, he published his *Recherches sur les causes des principaux faits physiques* in 1794; a *Réfutation de la théorie pneumatique*, which rejects the chemistry of his contemporaries; several *Mémoires de physique et d'histoire naturelle* in 1797. It is surprising to find him in 1798 interested in the influence of the moon on the earth's atmosphere. But already in 1798 the *Tableau encyclopédique et méthodique des trois règnes de la nature* was beginning to take shape, although it would reach its final form only in 1816. Then came the shells. While he continued to publish the meteorological annuaries from 1800 to 1810, while the *Discours d'ouverture* (introductory lectures) of his courses presented the progression of his broad general views, the *Système des animaux sans vertèbres* was published in 1802, along with a seemingly unexpected *Hydrogéologie*, all of his memoirs on fossils in the Paris area (1802–1806), and works on shells (1805–1809), to name only a few. In 1809 the *Philosophie zoologique* synthesized the Lamarckian perspective. Then follow more works on the determination of species, on "bloated polyparies" (*polypiers empâtés*)—*Penicullus, Flabellaria, Synoicum, Spongia,* on "corticiferous polyparies" (*polypiers corticifères*)—*Corallium, Melitoea, Cymosama, Antipatha, Coallina,* and so forth, until we reach the great *Histoire naturelle des animaux sans vertèbres* (1815–1822) and the last work, the anthropological crown of this impressive world view, the *Système analytique des connaissances positives de l'homme.*

If I have taken the reader through these works at a quick pace, without giving him a chance to catch his breath, it is because the dynamics of this production has much to tell us, if only we have ears to listen.

First of all, we have the impression that we see an immense landscape emerging, its essential lines seemingly buried in the variety of its features. But the structure gradually comes to the fore, while the diversity remains. Thus Lamarck's writings include studies and papers on perfectly delineated, minutely observed and analyzed topics. The frequency of the word "observation" (*observation*) is as noteworthy as the absence of the term "experiment" (*expérience*). In contrast, his work also includes the great synthesis, the system of living beings, which appears in 1801,[22] 1809,[23] 1815,[24] and 1820.[25] Synthesis and analysis are masterfully interwoven in the *Histoire naturelle des animaux sans vertèbres.* All of this raises questions. Why did Lamarck study all these disciplines in succession? How did he proceed from innumerable observations and classifications of living beings to the nearly total system of nature and of man? Are the connections that link the beginning to the end rigorous? Are there not one or more disjunctions? Each period of his life clearly corresponds to a specific field of inquiry; the disciplines of the preceding period then move into the background without being discarded, however. Thus he proceeded from meteorology and botany to biology, as his bibliography demonstrates; and biology represents both a culmination and a synthesis.

Should the question of the plurality of the disciplines he studied be resolved by "circumstances"?[26] At first sight, one might think so: If one of his fellow soldiers had not picked him up by the head for fun, there might have been a Marshal de Monet de Lamarck. Had he not been idle in the Provence, would he have made botanical fields trips? Had he not come to Paris for medical treatment, would he have attended the botany lectures of Bernard de Jussieu, while concurrently attending medical lectures, playing music, and observing the clouds because he lived as a pauper in a garret just below the roof? His acquaintance with Buffon, the institution of the *Jardin des Plantes*, the assignments he was given all impelled him even more decisively toward botany. Had the chair for animals "with white blood" not been the only one vacant, would he have taken up biology? So might one project the Lamarckian evolutionary model onto Lamarck. The development of his thought would have been to the encounters of his life what the fundamental "series" of living beings in the *Philosophie zoologique* is to the "circumstances" that have altered their "design" (in this case perhaps to be understood also in the teleological sense). Such an answer to the question overlooks the theoretical problem and does not sufficiently take into account his research, which first groped through a large number of interesting topics and then fastened onto living beings, without, however, removing them from their earthly, geological, mineral foundation and from the environment, which he explored right up to the moon itself! To be sure, our purpose is to raise the problem of transformism, not to describe in detail all aspects of Lamarck's work. At this point, it is nevertheless necessary to examine briefly their significance.

Following Edmond Perrier, one might well term Lamarck's meteorological works premature. Recently, Hans Reiff and Cornelius Schuurmans have written that the weather can be successfully forecast "three days in advance" and that before long this figure will be raised to ten. They speak of a "torrent of air" above us.[27] Today there are meteorological annuaries, offices, and centers. Lamarck's dreams have come true, for there was much dreaming in this paradoxically scientific project of establishing determinism in the clouds and imposing order upon them. The imagination,[28] which Lamarck considered to be the essence of genius, here served the purpose of scientific theorizing. In meteorology, it led the young observer to link changes in the clouds with the winds, and the latter with the atmosphere, which he considered a fluid tide like the oceanic tides. By analogy, his scientific imagination called into play the attractive forces of the moon on these atmospheric tides. Of course, Lamarck did not know about the properties of air (lightness, mobility, dilatability); he could not introduce measurements; he was not even considering the problem, for he was ignorant of his ignorance. In chemistry he was not well advised in his opposition to Priestley and Lavoisier. Their theories on the nourishment and properties of plants might have proved enlightening to him in botany. Lamarck realized that the inorganic order, although separate from the organic, is at the same time intimately connected to it. But he did not recognize the true modes of this relation in the development of the chemistry of his time. He did not neglect the role of physicochemistry in the operations of life, however, and he made extensive use of the concept of "fluid," inherited from the eighteenth century.

In botany, on the other hand, his work is very important. Even though the *Flore française* contributed no knowledge of new plants until the third edition (for which Auguste Pyrame de Candolle would bring him Swiss samples), his production was very fruitful from the theoretical point of view, for it shed light on the conceptual distinction between the natural and the artificial. To organize plants according to their real structures (a concern of Bernard and Laurent de Jussieu) and at the same time to give the public a convenient and sure way of identifying plants—such was the difficulty Lamarck elegantly resolved, thereby winning the bet that gave birth to the *Flore française*. He adopted a binary classification by genus and species. He invented the so-called dichotomous method, which "consists in confronting him who wishes to identify a plant with a succession of two opposite terms, between which he must choose before going to the next line, where he will find a narrower choice, until he finds the name of the plant at the end of the operation." This procedure was addressed to the nonspecialist. The first term in the

Lamarckian series is botanical, however; plants are organized into six levels: polypetaled, monopetaled, composite, incomplete, unilobed, and cryptogamous. The animal series also involves six gradations: quadrumana, birds, reptiles, fishes, insects, worms. The *Philosophie zoologique* and the *Système des animaux sans vertèbres* also adopt the idea of a series common to two kingdoms. We intend not to examine all the areas in which Lamarck took an interest, but rather to illustrate the kind of attention he devoted to all of the sciences. The fundamental idea is that of system, which implies a relation between all of the structures of reality. Lamarck may have erred completely with regard to the physicochemical world, but he saw that living beings are intimately united with it because their structure participates in the physical and chemical realms and because the environment (to use a modern term), "circumstances" or "external factors" (to use Lamarck's terminology) modify the organization, to the extent of transmitting themselves from the transformed organism to its descendants. From the moon to living beings, the economy of nature is understood as one, and it is this unity that Lamarck seeks to demonstrate in his work and in nature.

The two fields of geology and paleontology considerably enriched evolutionary biology after Lamarck, but fossil shells already played a role in his transformist arguments. With regard to the *Hydrogéologie*, published in 1802 at the author's expense, T. H. Huxley wrote that "the vast authority of Cuvier was employed in support of the traditionally respectable hypotheses of special creation and of catastrophism; and the wild speculations of the 'Discours sur les Révolutions de la Surface du Globe' were held to be models of scientific thinking, while the really much more sober and philosophical hypotheses of the 'Hydrogéologie' were scouted."[29] The immensity of time constitutes the dimension of the book, in those days an unimaginable dimension,[30] which Lamarck nevertheless imagines. Fresh water and salt water, the compostion and displacement of the ocean floor, the organic origin of all composites are the main objects of study through which uniformitarianism emerges, which during this period stands in opposition not to a dialectical vision, but to a theological ideology of "catastrophes."

The relation between water and land, and between each of these and fossils, is evident. If to this one adds the fact that fossils to a large extent suggested the notion of evolution and the true significance of living beings, the unique structure of the system appears once again in all of its clarity. We borrow from Bernard Mantoy the following quotation from the *Hydrogéologie:* "In the globe that we inhabit, everything is subjected to continuous and inevitable mutations that result from the essential order of things; these changes take place, indeed, with more or less rapidity or slowness, according to the nature of the structure of the objects; they are nevertheless taking place at any given moment."[31] We are far from Cuvier's six thousand years.

In the *Hydrogéologie*, one easily moves into paleontology—the two fields shed light upon each other. If the shells have completely retained their valves, it is because these animals lived where they were found. If we find coastal fossil shells, it is because no flood destroyed the shores. Such is the evidence that the fossil shells from Passy and other basins brought to Lamarck—at the same time it is both paleontological and geological evidence; but contrary to Napoleon's wishes, it does not confirm the Bible.

Thus "circumstances" played a role in Lamarck's life only to the extent that he was Lamarck, an impassioned observer fundamentally convinced of the sovereignty of system.

Who was Lamarck? I suppose the irreplaceable and unique Jean-Baptiste de Monet de Lamarck stands in that unapproachable space where the singularity of his being is tied to the profound penetration into the very substance of the historical structure he carried within himself.[32] Second, I suppose that it is difficult to use the word "genius" in an essay, not because it is meaningless, but because its content has not been seriously analyzed,[33] but has always been the object of denials or ideological incantations. Having made these reservations, we can now make two statements about Lamarck's social significance. If he seems to have been the toy of his circumstances, it is at the level of his birth and of the poverty that pursued him throughout his life. He belonged to the nobility that if not penniless, at least had few means—and for that very reason Lamarck was likely to fit into a politically revolutionary situation, even if he did not play a specifically political role. But his so-called fantasies were the result of his need to make the most of his encounters and to overcome obstacles in order to reach a more secure social position. From the Jesuits to the military career, from the military career to the small bureaucratic job, and on to the scientific position—such is the story of a man who was an heir in name only, and who therefore is not what today and in sociological terms we would call an heir. Moreover, the "Lamarck problem" becomes somewhat clearer by virtue of the constant, even posthumous struggle between Cuvier and Lamarck. But the notion of precursor does not come into play here. It is rather the boldness and the intellectual freedom of a man who disregards social and religious taboos. In his eulogy, Cuvier accused him of having risen thanks to his connections. The simple fact of the matter is that he was supported in his legitimate rise under Louis XVI as well as during the Revolution, and then was abandoned by the emperor and the Restoration. His independence from the Book of Genesis is striking in several passages of the *Hydrogéologie,* where the Flood is criticized. His concept of the Supreme Being has nothing to do with religious attributes. Lamarck's God is the originating principle of evolution. Cuvier's theological afterthoughts are unmasked, and he comes very close to being named. The nebulous myth of the precursor is beginning to fade, while there is seen to emerge Lamarck's by no means immature attempt to eliminate a number of ideological obstacles to the positive construction of the zoological philosophy, in other words, biological theory. His struggle against a radical separation of the species; his fight against catastrophes that would explain extinct species (in which he did not believe, as we shall see); his notion of extended time, which is the necessary and first field of transformations—all of this is the making of a science that would be truly scientific. To be the pioneer of this undertaking is not to be a precursor—it is to be in his own time someone who clears and opens a way at his own risk, peril, and personal damage. The courage and disinterestedness that the child on the battlefield of Willinghausen already possessed to a surprising degree did not fail the adult. He was deficient in neither observational acumen nor synthetic power.

This is how Lamarck ought to be seen; this is how he stands in the critical struggle of the eighteenth century and in the cultural effects of the French Revolution.

2 Nature

The end of the eighteenth century and the beginning of the nineteenth century is a period of disjunction between what had been the "natural history" dominated by the illustrious Buffon, and what would be called "biology" by Lamarck in France and Treviranus in Germany, among others. At that time, general considerations on the universe and nature formed a monumental whole to which the most brilliant pages of literature, science, and philosophy contributed. The "concept of nature" was the strategic point at which the obsession of a religious mystery of creation confronted the will to oppose a natural course to the supernatural. Nature appeared in turn as a sign, a delegate, and a rival of divine power.

Thinkers did not seem conscious very often (let alone necessarily) of the twofold movement in which they participated and the contradiction they embodied. Philosophies—whether spontaneous or deliberately constituted—found expression among the learned and scientists; scientific considerations were implied, either openly or implicitly, in the minds of philosophers as well as literary authors. Finally, all the currents of European thought converged on Weimar, where an extraordinary man lived, whose thought was capable of embracing and dominating these complexities. "Goethe who in his endeavors to investigate nature would willingly encompass the Great Whole . . . who is always upon the track of some great synthesis, but who from the want of knowledge of single facts lacks a confirmation of his presentiments, seizes upon, and with passion such decided love, every connection with important natural philosophers . . . He will always go further and further."[1]

After the great quarrel between Geoffroy Saint-Hilaire and Cuvier, Goethe wrote in 1830, "The best of it is that the synthetic manner of treating nature, introduced by Geoffrey [sic], cannot be kept back any more."[2] It is in this philosophy of nature that Lamarck's "zoological philosophy" fits, as he will call his famous 1809 work on transformism. Certainly Goethe did not quote Lamarck, nor Lamarck Goethe, but Geoffroy Saint-Hilaire also wrote his *Principes de philosophie zoologique* in 1830. Philosophy of nature—not to be confused with Schelling's *Naturphilosophie*—was in the air of Europe.[3] It fit into the movement of reaction against scientific empiricism, itself in opposition to the idealist philosophies, which to the scientists seemed so far removed from the facts. At least this was how Haeckel saw the situation in Germany. Even though the French were not dealing with the same set of questions, a similar craving for theory and unification could be found on either side of the border, in Treviranus or Oken, in Lamarck or Geoffroy Saint-Hilaire.

What does the term "philosophy" signify in conjunction with the adjective "zoological?" It expresses a demand for generalization, for systematization, for unification. It requires the subsumption of details, of the infinity of facts under principles. In a word, it expresses the need for a theory. Although ideology permeates this construction proposed as theory, the latter seeks to find itself, even through its aberrations. The analysis of the "organization" that transmitted the description of "visible forms" gives body to the notion of a "whole."[4] The genealogy of livings beings has a serial, hierarchical order, and the breadth of

Lamarckian wholes derives from the importance of the concept of "branched series" (*série rameuse*), which he first formulated in the *Discours de l' an VIII,* and which recurs in all the opening lectures of his courses. The unique and graduated "series" of living beings is the very order of nature. The knowledge of origins and the analysis of "natural" connections replace nomenclatures,[5] whose artificial character stands in opposition tò the "real" order of nature. At the beginning of the *Discours de l'an XI* (1803), referring to orders, genuses, and species, Lamarck advises his listeners never to forget "that all of these indispensable divisions are fictitious, and that nature recognizes none of them." Classification makes possible the accumulation of data.[6] "True objects" are not the result of classification; true objects are produced from principles and laws that constitute "the philosophy of nature." This is the true philosophy of natural history: "It is known that every science has or must have its philosophy; it is also known that every science achieves real progress only through its philosophy."[7] Nature cannot be confused with "art." Every analysis requires the whole. It is zoological philosophy that produces "biology," that is, the logic of life that the natural history of the eighteenth century had not discovered.

Lamarck's *Discours d'ouverture* are characterized by a remarkable power of thought and of expression. From the outset, it should be noted that if the author of this "zoological philosophy" could not do credit to his grandiose program, if the order in which he proceeded is questionable, he nevertheless established synthesis as a criterion of science; he made theory a precondition of biology. The *Philosophie zoologique,* the great book of 1809, was conceived as a body of "principles" and, by means of them, of "precepts," a set of general characteristics. Lamarck became with one stroke the founder of "biology"[8] and transformism. The inseparability of the biological order and the succession of increasingly complex living beings ought to be stressed. The zoological philosophy therefore presents the series of animals and, in the same breath, the logic of the biologist's steps and the logical connections between the results of these steps.

It is easy to understand why Lamarck placed the concept of "nature" at the very center of his "philosophical"—that is, "theoretical"—goal. Nature is indeed the most general idea, the keystone of all principles, which contains the essential structures that support the entire edifice. The time has now come to reflect upon the remarkable text in the sixth part of the introduction to the *Histoire naturelle des animaux sans vertèbres*. This reflection will not lead to a detailed textual analysis, for it would then be necessary to introduce the work. We shall nevertheless attempt to shed light on the struggle between the considerable ideological baggage of the idea of nature inherited from Diderot, d'Alembert, Rousseau, and, beyond d'Alembert, from Newton on the one hand and the theoretical function that is slowly freeing itself from the theological enterprise on the other. This struggle, this confrontation of contradictions is completely transparent in the text, to which we now turn.

"On nature, or on the power, in some sense mechanical, that has given being to the animals and made them necessarily what they are."[9] Thus reads the subtitle of this sixth section. One cannot marvel enough at the fact that the author could condense so many confrontations into three lines, that the ambiguous term "power" is made more precise by the adjective "mechanical," which is in turn tempered by "in some sense," which reintroduces a measure of mystery. The act of "giving being" discreetly suggests the Creator, but the adverb "necessarily," which corresponds to "mechanical" and excludes teleology, conjures up opposition to the idea of a divine person. We may also note that the terms "to give existence to" or "to produce"[10] make conspicuous the absence of the verb "to create." In spite of apparently precise definitions, nature is therefore the imprecise realm where the author of nature still hides, in the scientific emptiness clandestinely inhabited by a demythologizing intention. It is in this very realm that a complete rupture should have occurred, but nevertheless remained incomplete: the break between the necessity of laws and the divine will. The ideological order and the theoretical order remain locked in battle throughout this text; they make peace, only to pick up the fight again. Why did Lamarck not finish what he had begun? Why did he reach a point opposite the one he sought to reach, namely, the contamination of nature by the "Supreme Being," the author of the divine plan? "We shall examine . . . what this strange power might be, which is capable of giving existence to beings as admirable as the ones in question!"[11]

Lamarck's spellbound mood and enraptured tone betray on several occasions the ambiguity of the concept. But it would be a mistake to overlook the rhetorical skill that consists in giving way to fervor with respect to the "higher will," to the "intelligent and unbounded power," in order to make the reader admit later that God has two ways of acting, neither of which tarnishes the divine perfection. The first is the "immediate" and "simultaneous" creation of the species; the second is the instauration of a "special and dependent power that is, however, capable of giving existence to so many diverse beings," a "mode of execution of the supreme will" that is mediated by nature and ranges over the temporal dimension. It conforms to the divine majesty, but especially to knowledge of the facts. Lamarck thereby forestalls the scandal that might be unleashed if his thought were to be misunderstood as atheistic. After having taken great care to avoid this risk, he puts forward the hypothesis of the progressive genealogy of living beings through the brevity of the life of individuals and the relative but undeniable permanence of types. With the same skill, the founder of transformism enters into the logic of the creationist argument to the extent that he seems tempted to return to a direct consideration of the divine power as acting without the intermediary of nature. "Great indeed must be the power that was able to give being to all bodies, and to make them generally what they are."[12] Is nature sufficient to such a task? The answer is Yes, if the central power stands behind the delegated power.

One should not forget that Lamarck was granted honors during the French Revolution, at the time of the Convention. (Indeed, the use of the expression "Supreme Being" indicates a view of the eighteenth century seen through Robespierre. But one should also remember that Lamarck's career unfolded under the Empire and continued under the Restoration, at the beginning of which he wrote the *Histoire naturelle des animaux sans vertèbres*. This historical reminder introduces two remarks: First, different though they be, these three political milestones have one thing in common—respect for the divinity. It would not have been well advised to raise doubts about God, and Lamarck knew this, whatever his own intimate thoughts might have been. Second, while the tone of the text is undoubtedly religious, it is even more clearly juridicopolitical.[13] The central power acts in the most remote territories through the intermediary of powers delegated to the executive branch ("the mode of executing"). Ultimately, the creationist thesis is invoked and analyzed under three aspects: (1) direct creation; (2) simultaneous creation; (3) creation that makes fixed species emerge. The critique of creationism takes place at the level of the distinction between a divine power beyond the scope of questioning and modes of execution that can be brought into question. The *distinguo* is made possible by the introduction of juridicopolitical concepts. At the same time, the method of observation seems to be incompatible at the scientific level with the appeal to a God inaccessible in His very essence. Nature alone is accessible to observation. At this stage of the analysis, the natural is at its farthest remove from the supernatural; science is at the farthest remove from religion. The concept of nature is played off as a theoretical demand against the encroachments of theology; one sees the contradiction at work. The scientific enterprise progresses and brings about the transition from the concept of nature to transformism through the intermediary of the concept of "change." The appearance of new species, the modification of one species under the empire of "circumstances" testifies to the presence of a general power, "always acting, which changes, forms, destroys and ceaselessly renews the various bodies."[14] This power is limited, dependent, subject to the necessity of laws. And "this great power" establishes other powers according to the principles of good administrative organization. Here biology and politics become conjugates. It is a power "in some sense blind," "a power that, however great it might be, cannot do otherwise than what it does; in a word, a power that exists only through the will of a superior power without limits, which having established the former, is really the *author* of all that originates from it, of all that exists."[15] It is interesting to juxtapose these statements with those of François Jacob in *The Logic of Life*.[16] The general power that consecrates in its transcendent place the omnipotence of the sublime author dismisses teleology with the same stroke. Here Lamarck expresses himself as a mechanist; he would not even utter the word "finality," which Jacob uses, albeit in an exorcized form.

The text puts God and nature in their respective places. It does not substitute the one for the other; it does not break the chain of concepts, but rather orders them in a new way. More precisely, it brings about a transfer of God's power to nature; but this purely executive power does not include decision making, and therefore does not include setting goals. The execution is "mechanical." The subtitle adds the words "in some sense" because the divine plan transpires to some extent in the set of these causal chains in spite of this blind necessity. How could it be otherwise, since nature is realizing God's ends? Science, however, has no other task than that of bringing to light the means: "Now this circumscribed power, which we have scarcely examined, scarcely studied, this power to whose actions we nearly always attribute an intention and a goal, this power, finally, that always does necessarily the same things in the same circumstances and nevertheless does so many and such admirable ones, is what we call 'nature.' "[17]

At this point, one might think that all is clear. But if throughout this sixth part of the introduction the term "power" has defined nature, it is obvious that the latter is not only the set of the laws and movements of the universe. More precisely, nature suggests an activity, a dynamism that "in some sense" tempers the ruling mechanism. The opposition of the concepts of "nature" and "universe" exposes the distinction between the active and the passive, between what is foreign to matter and what is matter (the laws and motions). At the same time there emerges the risk of "personifying" this "set of active causes," this activity beyond matter whose auxiliaries are space and time. The universe, on the other hand, is nothing but "the whole of physical and passive beings."[18] Hence the awe that is difficult to repress in the presence of this "peculiar power that does so many things" and allows the teleology of the divine plan to transpire through the inverted language of mechanism.

Thus, in the hierarchy of being, the highest level belongs to the "will" of the "sublime Author" of nature. The second level is that of the strange power that is neither intelligence, nor matter: "It is neither a body, nor some specific being, nor a set of beings, nor a composite of passive objects; on the contrary it is . . . a particular order of things, constituting a real power that is nevertheless in subjection in all of its acts."[19] The definition of nature, or rather the most precise of its several definitions, reads, "Nature is an order of things extraneous to matter, ascertainable by the observation of bodies, and the whole of which constitutes a power inalterable in its essence, subject in all of its acts and constantly acting in all parts of the universe."[20]

The second level, the one that interests us here, must not be confused with the universe. We shall ask ourselves why Lamarck valued this definition so highly. The definition of the universe, the third level of being, reads as follows: "The universe is the set—inactive and without power of its own—of all physical and passive beings, that is, all existing matters and bodies."[21] It is completely extraneous to nature. However, it is in determining the changes undergone by bodies and matter, that is, by the universe, that the power of nature becomes manifest. The universe is to nature what nature is to God. Lamarck's analysis involves a double movement of separation and relation between these two pairs of terms. The universe is completely distinct from nature, just as nature cannot be confused with God, or "the watch with the watchmaker." Just as the divine plan is read in nature, just as it transmits to us something of God himself, so the power of nature is discovered in determining the characteristics, movements, and changes of the universe.

These confrontations bring to the fore the essential characteristics of each level of being. It is in relation to God that the subjection of nature to the necessity of laws contrasts with the complete freedom of God; but in the same relation one can also see the power and the activity with which nature is endowed and through which it presents a certain analogy to the divinity. In comparing nature and the universe, the activity of the one contrasts with the passivity of the other, the incorporeality of the one with the materiality of the other, the immutable character of the one with the transient character of the bodies that compose the other, even though the universe must last "as long as its sublime Author will allow it."[22] Nevertheless it is correct to say that the laws and movements of nature function within the spatiotemporal domain of matter. What is the place of life in this gradation of levels? The answer is complex, as the use of the expression "in some sense" indicates. We have seen that *life . . . in some sense* resembles "nature in that it is not matter but an order of things animated by movements, which also has its power, its faculties, and uses them necessarily so long as it exists."[23] Yet life differs from nature: Living beings have an ephemeral, transient character, whereas nature, which holds the key to their existence, is immutable and unalterable. Life thus stands as an intermediate suborder between the levels of the universe and of nature. Life is an invisible power dependent on the power of nature, destined to the destruction by which the living, in ceasing to live, once again becomes matter. Life participates in the strange singular character of nature, even as it is dependent on the inanimate. We are still in an era when life is thought to be a scientific concept whose science is biology.

Are these different orders, between which secondary mediations sometimes appear, tributaries of efficient or final causes? Finality can be postulated—but not penetrated—only at the level of God and the divine plan. Nothing in the universe or nature indicates the pursuit of an end. The necessity that reigns in nature is the opposite of finality: "neither intention, nor goal, nor will." Lamarck, who of course did not know about later development such as molecular biology or biochemistry, was nevertheless able to endow the mechanism with universality in spite of the naturalists of his day. But it is curious that he could justify this scientific view only with an ideological argument. If there seem to be foreseen ends, "it is because the diversity of circumstances, guided everywhere by constant laws originally combined for the end that the supreme author has in mind, this diversity of circumstances that existing things offer to him in every respect leads to products always in harmony with the laws that govern all the types of change it brings about."[24] Within the *Philosophie zoologique* itself, Lamarck relies on the physics of fluids, environments, climates, and so forth to justify mechanical efficient causes. But here, in this introduction from which final causes are banned, their appearance derives in the last resort from a divine finalism that cannot be deciphered in the natural causalities. It is a curious argument, which oscillates between the *Harmonies de la nature* by Bernardin de Saint-Pierre[25] just published in 1815 and the later mechanistic materialism of Ernst Haeckel. One understands why commentators hesitate between finalism and mechanism when they expound Lamarck. Indeed the latter saw clearly that science is tied to the exclusion of final causes.

Much later, in *Chance and Necessity,* J. Monod will see the exclusion of final causes as the principle of objectivity that supports all of science. Monod, who is no Lamarckian, emphasizes the apparent teleology of living beings and notes the perplexity that overtakes the scientist who confronts the skewing of mechanical principles in a single direction and the obligation under which the phenomenon places us to recognize "the telenomic character of living organisms,"[26] that is, the quasi-obvious pursuit of a design. Now the principle of objectivity that forbids us from interpreting nature in teleological terms also forbids us from adducing the slightest proof for or against the existence of God. This principle is nothing but a postulate. In a very different context, François Jacob writes, "The living being does indeed represent the execution of a plan, but not one conceived in any mind. It strives toward a goal, but not one chosen by any will."[27] This claim is not so dogmatic as it sounds—it only states that in the era of genetics the concept of a hereditary "program" made it possible to progress beyond "some of the contradictions that biology had summarized in a series of oppositions: finality and mechanism, necessity and contingency, stability and variation." In affirming the nonexistence of design, is F. Jacob taking an exclusively scientific position? The answer lies in the epigraph to his book: "Do you see this egg? With it you can overturn all the schools of theology, all the churches of the earth." This phrase from Diderot's *Entretien avec d'Alembert* unveils the ideological fallout and the presuppositions of will without choice, of design without forethought. J. Monod, on the contrary, in proclaiming the absence of proof either for or against finality, affirms by metonymy the absence of proof either for or against the existence of God. He thus places himself not inside, but outside the explanations of contemporary biology, and by this choice he betrays a certain fascination for what he himself calls the Kingdom. Both scientists live more than 150 years after Lamarck. They no longer have anything to do with ideas of "nature" and "Supreme Being." Nevertheless both men suggest—F. Jacob through Diderot, J. Monod through the anxiety about the meaning of life that he recognizes in man—that the "sublime" adversary is always reborn and that the death of God is more theological than scientific. Lamarck was in no sense decreeing the death of this God; on the contrary, he was establishing Him in such a place that science looks for Him only in introductions, preliminary considerations, and opening lectures!

But if nature can be explained, exposed only in mechanistic language, why call her a power? Indeed, why speak about her at all? Darwin would no longer need her. The "Supreme Being" could conceivably be hailed as a concession to social style. But nature? When Lamarck speaks of her, why does he show such great embarrassment and such a quaking of spirit? Why can one chase God from the scientific paradise only by ideologically overloading nature? At a time when the concept is inflated by a suspicious power, why is nature restored to a purely theoretical function with a simple and rigorous appearance? Why is a permanent contradiction manifest in the linguistic alternation of two categories of signifiers—those that gravitate about the metaphor of power and those expressed by the terms "order of things" or "whole?" The first series suggests personification; the second annihilates the first.

The one will never succeed in silencing the other. The disquieting question proves to be irreducible to silence. "This strange power," then later "This power must indeed be very great," and further "Can we thus deny the existence of a general power that is always acting?" and "Not only does this great power exist, but it also has that of instituting others" and again "What is nature? What is this singular power which does so many things and yet is limited to doing only those? What is this power?," "Again, what is nature? Might it be an intelligence?"[28] Enough of these quotations, these fragments of a leitmotiv that no definition succeeds in conceptualizing. Lamarck defines as much as he pleases, yet the question remains recalcitrant, for behind it there stands another: "Nature recognized attests herself of her author."[29] If nature witnesses to God, if the universe is the place where powers of change and the functioning of laws find their material support, could one not say that everything witnesses to God? It is indeed this relation of attestation—the language here is thoroughly legal—that refers the universe to nature and nature to God. The attestation is necessary precisely because there are three distinct realities. Pantheism is ruled out. "Many people assume a universal soul that directs toward a goal to be reached all the motions and changes that take place in the parts of the universe."[30] I do not think that Lamarck knew the *Ideas for a New Philosophy of Nature* by Schelling (1797) or the *World Soul, a Hypothesis of the Highest Physics to Explain the Universe* (1797) by the same author, or even his *Introduction to a First Sketch of a System of Philosophy of Nature* (1799). Lamarck was thinking of the classical humanities, along with certain common and almost popular notions. The man of the Convention and of the Empire rejected this pantheism, which the German principalities found perfectly tolerable. This rejection went hand in hand with that of "final causes." Nature has neither soul nor intention. The Supreme Author must therefore be at work, and it is because nature bears evidence of Him that she fascinates us.

How ought we understand Lamarck's uneasiness whenever he tries to approach the idea of nature? Is it some sort of mystical enthusiasm, of *Schwärmerei*? Does it belong to the visionary, or at least religious, order? Such is surely not his intent. He takes great care to distinguish the "realm of realities" accessible to observation—the starting point of a rational procedure—from the "realm of the imagination," which is nothing but that of fiction and "illusions of every sort."[31] Every piece of knowledge that does not grow "directly out of observation or consequences drawn from observed and ascertained facts is necessarily groundless, and therefore has no solidity."[32] The Lamarckian method is not encumbered by a master idea or experimentation; it is still very close to natural history, far removed from the procedures derived from the physiological work of a Claude Bernard, for example. Lamarck nevertheless presents himself as very suspicious of the seductions of the imaginary, and very careful to verify the origin of our knowledge at the level of the sensible, the only source of positive knowledge. His conception of knowledge derives from Condillac. His conception of the study of life derives from the great classification efforts, hence from lists of facts and from lengthy and precise observations. We have seen the enormous amount of work he accomplished in his studies of botany and of the invertebrates: "By judging and comparing simple ideas acquired by sensation, man obtains complex ideas of the first order; by comparing and judging two or more ideas of this order, he obtains others of a higher order; finally, with these, or with others that he joins to them, of whatever order they might be, he procures yet others and so on, almost indefinitely."[33]

Even the imagination uses sensible data, which it modifies in a fanciful, not a rigorous way. But like the "fictitious" ideas of Descartes, the models are always reality itself. The seductions of freedom lead most men away from the austere limits of real consequences. The small number of scientists contrasts with the great numbers of imaginative speculators; "*reason* always severe and inflexible,"[34] with the delights of the imagination. In all of this, Lamarck fits precisely into the sensualism of Condillac, and into the tradition of the Ideologues, in whose writings Condillac appears frequently. The medical method of Cabanis[35] is closer to the method of Linnaeus than to Claude Bernard's *Introduction to Experimental Medicine,* and Lamarck has a solid hold on the end of the eighteenth century. If it is indispensable to discuss Lamarck's method even in general terms, it is because the procedure he has in mind and his position as a son of the eighteenth century are so many pieces of information that shed light on his approach to nature. Let there be no mistake: The question "What is nature?" and the fascination that this "singular power" exercises on him constitute a rational procedure. These questions are perhaps the sort reason "could not avoid" even though they go beyond its power.[36] The questioning of nature does not "completely" transcend the power of reason, but it does stand at the limits of its force of penetration. Cabanis notes[37] that some general laws of nature are not pervaded by complete certainty, but rest instead on analogies. Sometimes one must be "satisfied with approximations."[38] Such is this nature that Van Helmont called "the order of God" and Cabanis, a "force"—a term close to that of "power." The concept of "nature" is therefore always a limiting notion, at the horizon of knowledge. If Lamarck considered his study of nature to be rational, we owe it to ourselves to discover how it fits into the whole of his scientific procedure and whether in this case one can describe its function as theoretical, in spite of its ideological baggage.

An answer to this question presupposes that one has gauged both the extent to which transformism had been introduced and the obstacles that its diffusion met. In the first half of the eighteenth century, Benoît de Maillet already had some idea of the transformation of one species into another (e.g., reptiles into birds). At first, Lamarck's mentor and protector Buffon seems to have accepted the equal antiquity of the various species. But later, in the *Histoire des minéraux,* he admitted the priority of marine animals over terrestrial ones, as de Maillet already thought. And in the *Époques de la Nature,* Buffon had no doubts about the late appearance of man. All of these claims were made in *considérations*, but were not turned into a system. The idea of interpreting the days in Genesis as lengthy periods nevertheless opened the way to Lamarckian transformism. It was Buffon's student who would build the overarching hypothesis; it was the student who would stumble on the obstacle of creationism and collide with the irreducible adversary, Cuvier.

Science had some brilliant hours under the Convention, even though the latter is usually thought to have merely defended the territory and harshly restored order within. Thanks to Lamarck's own efforts, the Museum d'histoire naturelle became a research center at the heart of the Louis XIII's *Jardin des Plantes*, at the heart of the *Jardin du Roi*, henceforth a dependency of the Convention. Great professors were busy revitalizing science there. In mineralogy, Haüy was formulating the laws of crystal formation; in botany, Jussieu was substituting his natural method for Linnaeus's systematics. Under the Convention, Lamarck published the second edition of the *Flore française*. Lacépède had taken over Buffon's old position. Geoffroy Saint-Hilaire was beginning his works of "anatomical philosophy." Thanks to the latter's efforts, Cuvier was named assistant in comparative anatomy at the Museum in 1795; he would take over Daubenton's chair at the College de France. He was on the road to honors. He would die with the prestigious title of *pair de France*. Creationist doctrine would be the rampart of dogma and would fulfill Napoleon's wish, if it is indeed true that he told Cuvier, "Above all, do not touch my Bible!" Cuvier was a very great scientist. Creationism is an ideology used by scientists, founded on the dogma of creation. But dogma claims to be certain and makes scientific truth depend on theological affirmations that interpret Revelation and the Scriptures (or better, makes the natural dependent upon the supernatural). Fixism, whether or not held by the scientist, is founded on the direct action of the Creator, Who at the beginning caused all species to emerge simultaneously, once and forever fixed. They can disappear in the cataclysms of the earth, but cannot be modified.

Now the point of departure of the transformist hypothesis was the difficulty Lamarck encountered in his scientific practice when in 1793 the Convention put him in charge of reclassifying the "inferior animals," that is, the invertebrates. After encountering problems in the separation of species, he broke with fixism and came to think that the latter are transformed one into another. It was therefore insofar as he observed and classified that he discovered the impossibility of an act of separation of the species, in other words, of an act of creation for each of them at the beginning of the beginning. For Lamarck, the very principle of transformism and the struggle against creationism were tied to this ancient method of observation from which he would never free himself, which he would never complement with experimentation.[39] The revision of the concept of species triggered all of his "philosophical zoology." I believe that this point has not been sufficiently weighed. As we have seen, the *Discours de l'an XI* relativized our divisions. In the name of good classification practice, this very notion took on a relative value and demanded a rigorous distinction between the natural and the artificial. It was then that Lamarck concluded that species were transformed one into another by circumstances. It was then that he noticed the influence of the exercise of functions and the transmission of acquired characters—at least he thought he noticed this. It was in the *Discours de l'an XI* that he raised the question, "What is a species?," and at the same time opposed creationism. Species could not have been created on the morning of Creation for they appeared through the effects of circumstances, as time unfolded. They could not be fixed since they were transformed by these circumstances with one species emerging from another. They could not each be the result of a special creation, since they were formed by transformation. It was therefore imperative to stop imposing rules on the Supreme Author and limiting His power. Only nature is subject to laws; we can read them in her. We then see in reality the dissolution of the separations that "art" had established; we see the groups melt into one another. The separations are the effects of our ignorance or the conveniences of our practice. "To facilitate study, it is useful to give the name of species to those [beings] that resemble one another and reproduce themselves alike."[40]

The idea of God does not exclude the mediation of nature: "No one dare say that this infinite power could not will what nature herself shows us that she willed."[41] Divine power acts through nature: "Ought I not recognize in this faculty of nature the execution of her sublime author's will?"[42] Creationist doctrine was incompatible with transformation in time and space; therefore it had to be eliminated, lest it hinder research or make it stray. If God had withdrawn in order that the history of living beings might be set into motion, if events occurred in this history so as to modify the very organization of living beings, this "productive" ("noncreative") power somehow had to be operative in them; this power was Nature. Thus the fundamental principle of transformism and the impossibility of admitting creationism derived together from the Lamarckian elaboration of the concept of nature: "To elaborate a concept is to vary its extension and comprehension, to generalize it by the incorporation of exceptional traits, to export it out of its area of origin, to take it as a model, or inversely to find a model for it, in short, progressively to confer upon it, by regulated transformations, the function of a form."[43]

If Lamarck did not think through in advance these steps outlined by Canguilhem, he spontaneously followed several of them with regard to the concept of species. The use he made of the idea of nature at the end of these operations was not so much a destruction of creationism as the consequence of its collapse. We can then take a new step and ask whether it was necessary to invoke an idea as obscure as the vicissitudes of the text of the introduction to the *Histoire naturelle des animaux sans vertèbres* suggests. Let us suppose for a moment that Lamarck had simply analyzed the order of biological forms by stopping at the level that he defined as "the universe"; would not his reasoning then have been more satisfactory? The answer is a categorical No, for transformism would then have been absorbed into atheism by the defenders of the theological apparatus. It was therefore necessary to dissociate the refutation of creationism from any appearance of sacrilegious negation of the Creator. Insofar as she is distinct from the universe, that is, insofar as she holds power, nature brings about this dissociation by being interpreted as God's executive. Hence this double result: (1) God is made indirectly present to His creation through the distinction between His creative act and the modality of His act. (2) The transformations of species are made possible by this productive activity, which nature holds from God. Finally, nature is necessary as a weapon against ideology, but it is precisely this antiideological function that inevitably leads Lamarck to meet his opponents on their own ground, namely, ideology. Nature alone could fill this role.

Cuvier did not overlook the problem of transformism. He came to it not from the problem of species, but from the relation between existing races and the fossils, as befits his paleontological activity. "Why, someone might ask, could the present races not be modifications produced by local circumstances and climatic change, and carried to an extreme difference by the long succession of years?"[44] His refusal to admit such a hypothesis rested explicitly on the impossibility, at his time, of pointing to intermediary species. We cannot know to what extent his refusal to examine the hypothesis was unscientific, but Cuvier's creationism is indubitable. If I may bring in a biographical datum, I would say that Cuvier, born in a region under Würtemberg hegemony, trained at the academic college of Stuttgart, was much more sensitive to Germanic influence and better informed about German philosophy of nature than Lamarck, who had been brought up in the tradition of Rousseau and came to the Museum at the time of Robespierre. In eighteenth-century France, nature was associated with the Divine Author, and not at all with immanentism or pantheism. Moreover, comparative anatomy was frequently more favorable to fixism than the natural history of species. However that may be, Cuvier was to remain inflexible. And in Lamarck, the concept of nature, loaded with strategic significance, would be worked by a double movement. First, stimulated by the results of scientific practice, the concept very quickly found itself in the throes of an ideological battle. Hence Lamarck's embarrassment and the ambiguity of his discourse in the text examined earlier. Transformist thought developed in the shadow of this equivocal situation. For this very reason, the principles that he would put into place to complete his system did not achieve the level of true conceptualization, since they were implicitly permeated by deistic importations. The theoretical rigor of his thought would be affected by it as well, as we shall see. The whole question revolved around the status of nature. It would not, indeed could not, be clear unless the term were used synonymously with "universe," thereby becoming completely neutral, as in Cuvier. But then the term is merely a sort of linguistic convenience to designate the "whole"; it is not the strategic weapon of a battle. By a strange inversion, to the extent that he thought he was making more precise and rational the concept of nature, Lamarck was building stone by stone the monument of a "philosophical"—not a "zoological"—philosophy, a monument in which he would enshrine his biology.

If this is the case, what differentiates the philosopher's "nature" from that of the scientist, or from that of the poet? We shall illustrate this by comparing Schelling, Lamarck, and Goethe. It does not matter that Goethe paid attention only to Geoffroy Saint-Hilaire, and not to Lamarck, or that Lamarck ignored Germany, whereas Cuvier knew the German language and culture thoroughly. At the end of the eighteenth century and the beginning of the nineteenth century, "nature" was an overexploited idea. The philosophers of nature were dominated by the great figure of Goethe. Schelling, who was raised on Goethe, was protesting against the philosophy of Fichte, partly because the latter did not give nature a life of its own. The philosophy of nature reintroduced monism[45] by turning nature into "the visible spirit, spirit into the invisible nature."[46]

From the very outset, it is clear that the definition is metaphysical, and more specifically, pantheistic. The question "What is nature?" was not raised on the basis of its manifestations (as it was in Lamarck), but from a spiritual intuition (reflective even though intuitive). We are at the level of an interiority that reveals the unity of the world. We are far from the dualism of Creator and Creation, or of Supreme Author and Nature. Nature is a "spiritual power" that one grasps in the creative act of the spirit; it is not a delegated and executive power. The world has a soul, the *Weltseele;* we know what Lamarck thought of this.

In July 1798, after Goethe had called him to Jena as an extraordinary professor, Schelling would encounter art and poetry on the way, in Dresden; at the same time he would also encounter love in the person of Caroline Schlegel, who later became his wife. "The hour of the philosophy of nature had rung."[47] From Hegel to Hoelderlin, from Schelling to Novalis, an intimate relation is established between poetry and philosophy. This metaphysics is nevertheless impregnated with physics. It draws its material from the scientific theories and discoveries of the eighteenth and nineteenth centuries, but it takes only the matter, not the form of the procedure. In a strict sense, science illustrates "the activity of the infinite production of nature," on which Schelling insists in his *First Sketch of a System of Natural Philosophy* (1799), which led to evolutionary dynamism (in the universal, not the biological, sense of the word). We have considered this metaphysics briefly in order to measure its distance from Lamarck's writings. Lamarck is perhaps a scientist who strays; but he is not a metaphysician.

Goethe is, on the one hand, a man involved in scientific practice as an amateur of genius;[48] on the other, a very great poet. One crucial point unites the two aspects of Goethe—the idea of nature as "whole." In his case, it leads him to the resolute adoption of the principle of synthesis (Geoffroy Saint-Hilaire) against the excess of analysis of detail (Cuvier).[49] It is known that on the eve of his death, he discussed the famous quarrel between Cuvier and Geoffroy Saint-Hilaire in 1830. In his *Discours,* Lamarck had recommended that one begin with the whole. Even if Goethe unjustly ignores Lamarck, there is an undeniable convergence there. "Goethe . . . would willingly encompass the Great Whole."[50] This whole is for him the object of a vision, a contemplation of the great general laws. This contemplation easily leads to the admiration of natural harmonies. "Nature proceeds with such wisdom and moderation that a bird during its moulting never loses so many feathers at once as to render it incapable of flying sufficiently to reach its food."[51] When discussing cuckoos, warblers, and birds in general, Goethe's enthusiasm is boundless, and emotion then overcomes thought by substituting lyricism and even religiosity for rationality: "There is certainly something divine in this . . . which creates in me a pleasing sense of wonder. If it were a fact that this feeding by strangers[52] was a universal law, it would unravel many enigmas, and would say with certainty that God pities the deserted young ravens that call upon Him."[53] This does not mean that he allows an external teleology in the manner of Bernardin de Saint-Pierre. But "nature and we men are all so penetrated by the Divine, that it holds us; that we live, move, and have our being in it."[54] It is easy to measure what separates the expression of this religious and poetic sensitivity from Lamarck's rational and juridical construction. Here outpourings of emotion submerge thought; in Lamarck, questioning never departs from the style of the intellect.

Yet it might be possible to discover a place where these three figures could find each other. It would not be at the level of the methodical production of ideas, concepts, mastered procedures. Instead, it would be some high place like the hills that dominate Weimar, where Goethe observed plants and birds with Eckermann, the place in the mind's eye where these three procedures, still distinct, would all end up. Goethe, at the crossroads of philosophical, scientific, and poetic roads, expressed well this vision of transformism by considering a small snake: "What beautiful intelligent eyes! . . . This head announced many things, the miserable rings of this clumsy body have halted everything on the way. To this organization produced completely in length, nature has remained a debtor of hands and feet, and yet this head and these eyes surely deserved them. She often acts in this fashion, but what she has abandoned, she develops later, when the conditions become more favorable."[55]

3 *The "Series" and "Circumstances"*

One of the most important considerations of interest to zoological philosophy concerns the degradation and simplification that one observes in the organization of animals, in moving from one end of the animal chain to the other, from the most perfect to the most simply organized.[1]
Lamarck, *Philosophie zoologique*

A few lines later, Lamarck calls this chain a "general series," further yet "an animal scale." Is this synthesis a logical distribution or a chronological ascent? Is it genealogical? Nothing in this formulation transcends the logic of the natural distribution of living beings. To reverse the series, to start from the simplest would be a necessary but not a sufficient condition for speaking about successive productions of nature. But even a successful passage from gradation to production would still not constitute descent. In fact, Lamarck inverts the series as soon as he turns it into a principle. In the *Discours de l'an VIII* (1800), where the series is presented as principle of the natural order, he discusses at the same time the animals "with which nature began." Lamarck is mistaken in assimilating from the outset the simplest with the "beginning." In the *Philosophie zoologique,* chapter 6, sometimes "degradation," sometimes "progression [is at issue] in the composition of organization."[2] Lamarck never states clearly whether this progression is a succession or a simple logical relation. I nevertheless think that Edmond Perrier is correct to see the fundamental significance of the series in its epistemological value, in its opposition to artificial classification in the name of the natural order, that is, precisely in the name of "organization," the key concept in the transition from the eighteenth century to the nineteenth century.[3]

Around 1800 Linnaeus's method was still considered virtually untouchable; it had put an end to collections based on similarity. Analysis progressed to the elements; classifications were expressed by nomenclatures, while continuous linear "series" like the famous ones of Charles Bonnet were born. Yet even though men like Buffon could see that only individuals correspond to reality, these series took the species as an invariant foundation.

Now Lamarck's series abolished the primacy of the species; classification was only a convenient tool. In the table of the *Discours de l'an X* (1802), Lamarck distinguished twelve groups or classes, separated into vertebrates (mammals, birds, reptiles, and fishes) and invertebrates (mollusks, annelides, crustaceans, arachnids, insects, worms, radiaries, and polyps). The *Philosophie zoologique* listed fourteen classes:[4] Lamarck added cirripedes and separated the infusorians from the polyps. A few years later, Cuvier's classification would be simple, structural, and would lead to comparative anatomy. It would include four general "plans"[5]—vertebrates, mollusks, articulates, and zoophytes—common to fossil as well as extant species. Lamarck's series could be considered implicitly evolutionary, if degradation is taken for a sort of evolution, which must be inverted in order to intersect with reality. But the true evolutionary modes have yet to be born.

The genesis of the Lamarckian series emerges in the *Discours d'ouverture (Introductory Lectures)* of his yearly courses of the Museum. In the *Discours de l'an VIII* (1800), which barely antedates the publication of the *Système des animaux sans vertèbres,* the immense series of nature's productions is invoked as a principle of the natural order in contrast to artificial nomenclatures. The *Discours de l'an X* (1802) continues this reflection on the "astonishing degradation"[6] or again this "series of masses" that forms a chain. And alongside of degradation, one always finds the chronology of productions.

One might think that Lamarck could just as well have tried to establish a series, the principle of the natural order, by beginning with the simplest animals. But this suggestion overlooks the necessity of starting from a criterion, namely, the most perfect and best-known species, the vertebrates. The most perfect vertebrate is man; the place that each great group occupies is a function of its distance from the perfected human organism. Such is the first justification of the series. The second justification or "proof" relies on the comparison of two extremes, the most perfect and the least perfect, whose relation then serves as a reference. In the *Discours de l'an XI* (1803), the scientific function of the Lamarckian series is at its most untrammelled, freed from a concept of species tied to nomenclature. Indeed, nomenclature is declared unworthy of the scientist, and the series is opposed to it as being alone capable of revealing "the admirable and constant progress *(marche)* of nature."[7] Lamarck does not abandon the distinction of classes, orders, genuses, and species; but he warns his students against confusing a convenient tool with the reality of nature: "Never forget that all of these indispensable divisions are fictitious and that nature recognizes none of them."[8] We shall soon see that species have no place in the series. The concept of "mass" is the only unit allowed; it is an important unit of organization founded both on intraorganic relations and on relations among organisms. Since organization is recognized as the essential characteristic of living beings, nature's great divide runs between the organic and the inorganic. This distinction is infinitely more important than that of the three kingdoms. The concept of "series" has meaning only within the context of the following set of themes: the opposition of the natural to the artificial, from which the relativization of the concept of species derives; the establishment of organization as the sign of life; the constitution of a biological object that grounds the science called "biology" and has as its consequence the primacy of the organic-inorganic dichotomy. Such was the new emerging structure that Lamarck saw clearly. In my opinion, this is as important as the transformist significance of the inversion of the series. Here Lamarck stands in continuity with the scientific conquests of the nineteenth century. The term organization recurs frequently on any given page of the *Discours* or the *Philosophie zoologique.* A few quotations illustrate this brief analysis: "The series that constitutes the animal scale resides in the distribution not of individuals and species, but of masses";[9] "It is systems that become degraded" (in 1806 he calls them systems of organization); "The species change"—a short formula which anticipates transformism by affirming change in time and the absence of an absolute separation in space, so that species do indeed become an artifact of nomenclature; "To facilitate study, it is useful to give the name of species to those [beings] that resemble each other and reproduce themselves alike."[10] Thus the genesis of the concept of series throughout the various *Discours* is grounded in epistemology, even though the *Philosophie zoologique* draws from it the transformism already sketched in the *Discours.* It is then that Lamarck confronts fixism, which implies creationism. It was not Lamarck, however, but his opponents who would carry the debate onto the terrain of ideology.

At the turn of the eighteenth and nineteenth centuries "a new science was to appear, whose aim was no longer to classify organisms, but to study the processes of life. Its object of investigation was no longer visible structure, but organization."[11] François Jacob, as is known, organizes the Lamarckian period around this concept, which obsesses Lamarck and recurs everywhere in his writings. He discusses Lamarck extensively without

Table of the Animal Kingdom: The "series" in Lamarck's *Discours*, showing the progressive degradation of the special organs down to their annihilation[a]

1 **Mammals**	Viviparous, with mammary glands; four jointed members dependent from the skeleton; lungs; hair on some part of the body	A spinal column as the base of a jointed skeleton	Annihilation of the special organs
2 **Birds**	Oviparous, no mammary glands; four jointed members dependent from the skeleton; lungs; feathers on the skin	Two-ventricle heart, warm blood; a brain and nerves	No more mammary glands; no longer a complete diaphragm
3 **Reptiles**	Oviparous, no mammary glands; four, two, or no members dependent on the skeleton; lungs at all times or only at maturity; neither hair nor feathers on the skin		
4 **Fishes**	Oviparous, no mammary glands; gills at all times, or at least before maturity; fins; neither hair nor feathers on the skin	Single-ventricle heart, cold blood; a brain and nerves	Incomplete and degraded skeleton; no more arms dependent from the skeleton; no more larynx, no more voice
5 **Mollusks**	Oviparous with a soft body having neither joints nor rings, with a variable coat; gills	No spinal column; no real skeleton	No more eyelids; no more lungs; no longer a true skeleton
6 **Annelids**	Oviparous with a soft body, extended, with rings, without jointed legs; no metamorphosis; gills	A brain in some, extended marrow in others; arteries and veins	
7 **Crustaceans**	Oviparous, with jointed body and members, with crustaceous skin; no metamorphosis; gills		No more heart

8 **Arachnids**	Oviparous, with jointed legs at all times, eyes in the head; no metamorphosis; spiracles and tracheae		No more arteries or veins; no more conglomerated glands for secretions
9 **Insects**	Oviparous, undergo metamorphoses; in their perfect state, have eyes in their heads, six jointed legs, spiracles, and tracheae	A longitudinal marrow and nerves; neither arteries nor veins	
10 **Worms**	Gemmoviparous, with a soft body, regenerative; no metamorphoses; never any eyes, or jointed legs; spiracles		No more eyes; no more tongue; no determinable sex
11 **Radiaries**	Gemmoviparous, with a regenerative body without eyes, or jointed legs, having a radial disposition of their parts; aquiferous tracheae		No longer a head
12 **Polyps**	Gemmiparous and fissiparous, with a body that is almost everywhere gelatinous; regenerative, having no internal organ other than an intestinal canal with one opening; in the last order, which ends with the genus *monad*, every special organ is annihilated and generation is merely *fissiparous*	No special organ for feeling or for circulation; never a head	No longer a special organ for feeling or for the motion of fluids; no regenerating gemma; annihilation of every specific organ

[a] The progression of degradation is nowhere regular or proportional; but it obviously exists throughout the whole.

adopting the leitmotiv of the precursor, but by locating him precisely and exactly in the "present" of the history of science.

What does the series look like? The table of the *Discours de l'an X* is reproduced here. If we read it from left to right, the first column contains what Lamarck terms the classes or great families, for which he introduces the term "masses." He was concerned not to fall back onto one of the categories of the nomenclature, but to use a term foreign to classifications to set up "important groups whose general organization depends on a given system of essential organs."[12] In 1802 and 1806 he put it as follows: "By masses of animals, I mean natural classes and great families, that is, the great recognizable portions of the order of nature." Already in the *Recherches sur l'organisation des corps vivants,* he had referred to "a general series by means of the masses" (1802). Internal organization, not visible characteristics, was to determine the rank in the series. Organization is fundamental and less fragile. Nevertheless, in the second column, alongside the lungs and the skeleton, one reads "hair on some part of the body" or "neither hair nor feathers on the skin." Thus in practice, Lamarck sometimes oscillated between essential and secondary features. But what counted above all in the identification of the "masses" were the internal relations.

Now one might expect the system of essential organs to be emphasized, the relations to be structured for each of the twelve masses in the left-hand column, the system to be perfectly distinct from the visible form, and the order to be governed by what François Jacob calls "functioning," that is, the connection between organs and functions. Could Lamarck have done that? François Jacob notes that the eighteenth century had neither "the concepts nor the technical means to investigate the hidden structure it [had] postulated."[13] Under these conditions, the series is limited to the most narrow range of possibilities and anticipates nothing in the future. And when Jacob remarks a few pages later that what count are not only the organs, "but also the way in which they are related to each other, . . . their organization,"[14] he seems to shed light on Lamarck, even though he is merely characterizing a period. The striking feature about the first two columns is the striving toward unification and the paucity of means for attaining it. Lamarck was looking for wholes; he knew that a scientist could not merely gather details. "Starting with each of the classes, I shall show that they are based upon the consideration of organization." In my opinion, he did not show this. The homologies and analogies of Geoffroy Saint-Hilaire and Cuvier are missing here;[15] they would have allowed one to see and to demonstrate the plan of organization. In this sense, the continuous linear series fails. The third column establishes divisions within the vertebrates and the invertebrates. Their system of essential organs is condensed, but even here one finds alongside the levels of marrow, arteries, and so forth the unexpected remark "never a head" in reference to polyps. The last column lists the "annihilation of special organs." The sequence of degradations does not clear the way for an intelligible law. The successive disappearances are announced by the expression "no more . . . , no more. . . ." One need only substitute "not yet . . . , not yet . . ." in an inverted reading to stand at the threshold of transformism. The table thus remains a function of a "knowing" closely related to "being," but not a function of being in the process of becoming, of productions of nature in time.

The concept of series is imperfectly realized not merely because the lack of necessary concepts makes it unrealizable or because it draws attention to a problem more than it finds its solution. The concept is called in question at its very core. If "we are not masters of the series," if we only have "portions of series," then the missing parts may be decisive for our understanding of the whole. But lacunae and our ignorance are not the only problems. The series has failed, confused by a proliferation of secondary forms, so that we do not know whether we are dealing with products of the primeval command or modifications produced by "circumstances." The series is "branched" *(rameuse)*.

Now the notion of "circumstance" enters the picture. What are these "circumstances"? Everything is happening as if two opposite factors worked together in the order (or perhaps the disorder) of production of animal forms; they hamper each other, only to come together at last in a whole where the power of the series finally makes its way with great difficulty through obstacles it overcomes or bypasses. These obstacles are the circumstances. They are responsible for the "irregularities" or "anomalies" or "deviations." It is at their level that the question of adaptation will be raised.[16] A given organ should have disappeared, but did not; sometimes one that has disappeared reappears. Forms proliferate as the result of circumstances. What does this mean? There is no definition. We have attempted to follow the development of this term throughout the *Discours*. In that of Year X (1802), it is presented as the source of the acquired characters and the species that proliferate around the masses. (The same expression moreover refers to the effect of the circumstances and to the species.) With the concept of "circumstance," we have progressed to the chronological and transformism. For circumstances are the spatial environment, but also that which occurs in time. A history has been retraced. We have followed its developments. Ordering is not the issue. The *Discours de l'an XI* contains the famous analysis of species where it is said that "every organization and every form acquired by this order of things and by the circumstances that work to bring it about has been preserved and transmitted successively throughout the generations until new modifications of these organizations and forms had been obtained by the same route and by circumstances."[17]

Although the latter are not defined, a few are enumerated: time, place, climate, and so forth. We can try to make the concept more precise by considering its etymology: "that which stands around. . . ." We shall then have an equivalent in the external environments *(milieux)*—not *the* environment *(milieu)*,[18] a term which is no more Lamarckian than adaptation. Although it is necessary to recognize this still nameless reality, the terminological blank raises a theoretical problem. Furthermore, when confronted with the mysterious serial power and its compelling intelligible necessity, the contingency of circumstances becomes dependent on the blind game of determinisms; we would not say chance, but chances. The mode of influence of circumstances on the organism would pass through mediations: they provoke needs, which provoke habits and actions that modify the organs through use or disuse. The *Discours de l'an X* alludes to examples that recur in the *Philosophie zoologique*, and that great biologists of a later era adopt as illustrations of adaptation without discussing them, for example, the blind mole, the aspalax, birds with webbed feet, and the long neck of the giraffe. The series is the norm; these examples, the anomalies. Strange causes have disturbed the regularity of an unfolding that had not been made to be disturbed.

And here we are surprised: After starting from a scientific perspective, we suddenly find ourselves in the midst of questions with a strong ideological flavor. From what power does the series emanate, if not from nature, which is so often invoked by Lamarck?[19] Are circumstances adversaries of this *natura naturans* ("nature naturing")? They themselves are of nature, which is diverse and changing; in its characteristics resides the cause of circumstances. But this is neither an answer nor an explanation.

Worse yet, by analyzing the circumstances, we have returned to the question of the very being of the series, which we had raised earlier. What would the series have been, if it had not been disturbed? Lamarck's table is the table of an altered series, not that of the original series. It reflects a lost purity and a universe of mixtures. Thus in Bergson the world of the species is the product of a mixture: that of the vital impulse *(élan vital)* confronting matter.

At what point in the logic of the argument, and why, did we move from the scientific to the ideological? At every stage of his thought, Lamarck has appealed simultaneously to the pure scientific causality of nature and to "first author of all things." Thus in the *Discours de l'an XI,*[20] within a few lines, he appeals to the first author and his infinite power; the prodigies of nature, which "the sublime Author must have willed that she possess"; and finally the necessity of including in his "biology" this "natural power" that represents the Supreme Being. The word "design" applied to the totality of living beings points to the author of the design. At every instant the passage from the scientific to the ideological is latent in a *philosophical zoology* of this sort. Lamarck cannot do without this "mass of principles and laws that constitute the philosophy of nature,"[21] and he adds, "This is the true philosophy of natural history, and, as is well known, every science has or must have its philosophy." It is less well known that "every science progresses only through its philosophy."[22] The series is the great principle of this philosophy. At the same time it is, according to Lamarck, a "positive fact," the product of a "constant law of nature."

Are the circumstances as easy to recognize, as easy to observe, at least indirectly, as the facts that result from them? *Rananculacea* have a different form and a different name, depending on whether they develop in air or in water. The anomaly lies at the level of the plan's execution, as breeders make us aware. We literally see varieties, if not species, in the process of formation. According to Lamarck, to be convinced of the unquestionable regularity of the series in spite of its anomalies, we need only "look" at the facts. But as always, Lamarck invites us to consider wholes from afar, then to come closer to "observe" the details without stopping our study there. Moreover, he "shows" us the details in a separate study of the organization of each mass. This study takes up the end of chapter 6 in the *Philosophie zoologique.* Our gaze moves constantly from facts to inductions that transcend them. If as readers we suddenly find ourselves in a certain ideology of the Supreme Being and of His delegate, nature, it is because Lamarck is always working simultaneously at several different levels. When he appeals to one of them, the others are always implicitly present. He is an instance of the scientific thinker who cannot succeed in breaking with the dominant ideology of his early years. Lamarck is not a precursor.

The exploration of the possibilities of a period, and careful attention to the concepts that are used in it as well as to those that are not but that we are tempted to insert retrospectively, makes it possible to avoid myths and to come closer to reality. At the time of Lamarck, this reality is the solidarity between the concepts of organization, biology, transformation,[23] and time.

To connect these ideas in the process of conceptualization, it would be necessary to begin with "circumstance," which brings about "change," which in turn requires an immense "time" to reach the present diversity of living beings. The foundation of transformism is not so much the original series as the circumstance that stimulates the "change," which in turn presupposes the "series." Since life is born from life, the earth is now populated by many varied organisms only as a result of successive reproductions, as François Jacob has written.[24] Time and reproduction are intimately connected. The momentous question of heredity will be discussed in chapter 4.

The idea of transformation stands at the very heart of Lamarck's system. But this system also requires "an impulse derived from organisms themselves, leading them gradually from the simple to the complex through terrestrial vicissitudes. It is the product of an ever unstable equilibrium between living forms. It is a network of interactions between organisms and their environment. It is the dialectic of similarity and difference in a unified history of nature. In short, transformism is a causal theory that accounts for the appearance, variety, and kinship of species."[25] The anomalies are simultaneously obstacles to the production of the original series and the foundation of the production of the world's diversity as perceived by man.

In fact, none of those who are called predecessors (Diderot, de Maillet, Charles Bonnet, Buffon, or Maupertuis) reached a similar holistic construction. Surprisingly, Lamarck reached it circuitously, by way of the degenerate series, which is closely related to the critique of the concept of species. "A system of relations between the constituents of a living being was not necessarily immutable. It could be transformed into another system one degree more complex, by a one-way process," writes François Jacob.[26] Lamarck had generalized his model. In the *Hydrogéologie* (1802), his concept of time was already broad enough to structure the history of living beings. Cuvier's time seemed to him ridiculously small. Cuvier did not need endless succession for a fixed world. Lamarck was apparently setting up the great ideas that would make good in the nineteenth century. But he was also apparently unable to give them a completely scientific status. Unless I am mistaken, François Jacob sees in him the intermediary between two centuries. In this regard, Lamarck's concept of series is quite remarkable: On the one hand, it allows him to become aware of the transformist idea; on the other, insofar as it is linear and continuous, it stands in a certain sense in the tradition of the earlier "scales" or "chains."[27] The concept therefore implies both a conceptualizing direction and an ideological survival.

This ideological density can be penetrated only if the series is linked to the idea of nature. One might even wonder whether this "plan" can be conceived without teleology. But it is neither the teleology of Bernardin de Saint-Pierre's *Harmonies naturelles*; nor that of Linnaeus before him, who saw the divine order in the deciphering of codes; nor that of William Paley, whom Darwin would oppose.

What then is the power that effects this constant, gradual increase in perfection? Lamarck rejects final causes: thus he appeals to mechanical causality whenever he sets up an argument. A hidden teleology is translated into a plurality of efficient causes forming a whole. It is a natural theology, however secularized it might be, that undergirds the edifice. This ordering of causes translates the ideological-scientific ambiguity that we see as the essential trait of Lamarck's production. But, someone will say, it is characteristic of every scientist that he has a spontaneous philosophy, an "SPS,"[28] and a nonspontaneous philosophy. Certainly, but in this case, we are dealing less with Lamarck himself than with biological science. The psychological ambiguity points to an epistemological ambiguity—the ambiguity of a science in the process of being born.

The ideological character of the series is not new. Its history has been written. There is no need to write it again or to summarize it—we shall only establish the specificity of Lamarck's series with respect to those that preceded it. With good reason, Daudin[29] traces back to Antiquity this concept, "flexible in its formulations" and in "its development," of a scale of animate beings. The world of living beings is represented there as "a set of forms ranked according to a regular and almost insensible degeneration from a maximum to a minimum of vitality." The great ancestor is Aristotle. His entire philosophy is a scale of being, where form and matter constitute a hierarchy in which every form is matter for the form that stands above it. At the base lies a matter that is the form of nothing; at the summit stands a form that is the matter of nothing, since it forms and moves itself. But in fact, Aristotle has taken care to specify the hierarchy of living beings in his *History of the Animals:* "Nature proceeds little by little from things lifeless[30] to animal life, in such a way that it is impossible to determine the exact line of demarcation or on which side thereof an intermediate form should lie."[31] The animal and vegetable series follow each other on a single line. The hierarchy of faculties of the soul determines the hierarchy of living beings, which is itself dependent on the great categories of being: potentiality and act, matter and form. This powerful synthesis dominated the following centuries. Leibniz himself would refer to it.[32] Starting with the soul as a "first entelechy" and "primitive moving force," Leibniz develops a dynamic conception of nature against the mechanism of Sturm, beginning with principles that make possible the analysis of a continuous process of living beings, which he did not formulate in his philosophy.[33]

This synthesis influenced the eighteenth century, especially scientists like Maupertuis and Charles Bonnet, but also all the naturalists Lamarck knew or followed, particularly Buffon, Jussieu, and Adanson. From the series, it is a short step to descent. Linnaeus himself had not rejected this natural order. But Lamarck used the series in a way that none of his predecessors had ever dared. In Daudin's words, "Adanson had first postulated families, whose unity he thought was guaranteed by a complete concordance of all determining characteristics, and then attempted to connect them according to the majority of their resemblances. Lamarck was much bolder—he thought that not only the succession of families among themselves but also the extent of each and the arrangement of their genuses could emerge exactly only from the total constitution of the series."[34]

Lamarck's series is farther removed from Charles Bonnet than from Adanson. Michel Foucault has characterized Bonnet's scale as "a subjection to time of living beings that had been considered simultaneous,"[35] not in the sense that the series would give birth successively to groups spread out in space in order to classify them, but in the sense that the totality of beings would move globally "on the time coordinate," to use François Jacob's expression. In this movement, beings and groups would retain their mutual relations. Time measures this movement toward the perfection of God, to which man would approach by leaving to inferior species the space he presently occupies. A day might come when Lamarcks and Cuviers might be beavers, or conversely. This is "palingenesis" in the proper sense of term.[36] The entire hierarchy is swept away in a metamorphosis that is a function of a temporal displacement. As Michel Foucault puts it, this is not an evolutionism, "but a *Taxinomia* that encompasses time. A generalized classification."[37]

François Jacob has compared the transformations of de Maillet in *Telliamed*,[38] the progression toward complexity in Robinet's *De la nature*, and the translation of the living world of Charles Bonnet in *La Palingénésie philosophique* and *La Contemplation de la nature*. De Maillet focuses on the transition of certain aquatic animals to terrestrial life and does not deal with increasing perfection as a function of time. Robinet gradually establishes the diversity of forms by "combinations and variations of the prototype." A prototype is a sort of "organic molecule" from which living beings emerge, but without passing from one form to another. Bonnet's scale is a construct deeply imbued with religious fervor and lyricism. He celebrates the harmony of nature and attributes it to the "adorable hand" in a manner that brings to mind some passages from Linnaeus. Sometimes this enthusiasm is transferred from God to nature. Thus in the sixth part of the *Palingénésie* we read, "The universe is in some sense of a piece. It is one in the most philosophical sense. The artificer has therefore formed it with one stroke."[39] The expression "natural harmony" recurs frequently: "We do not know where the organization ends or what its smallest term is. But in ceasing to organize, nature does not cease to order or to arrange. It seems that she is organizing even when she is no longer organizing."[40] The tone is very close to Goethe. To ensure unity, it is necessary in theory, if not in fact, to fill the empty spaces with "contiguous beings." "There is no leap in nature; everything is graduated and nuanced."[41] The only empty spaces are those of our ignorance. "The polyp links the vegetable to the animal; the flying squirrel unites the bird with the quadruped. The monkey touches the quadruped and man."[42] This unity comes from "the same general design," which embraces all parts of earthly creation: "Between the lowest degree and the highest degree of spititual and bodily Perfection, there is an almost infinite number of intermediate degrees. The succession of these degrees constitutes the universal chain. It unites all beings, links all worlds, embraces all spheres. One Being alone stands outside this chain: He Who made it."[43] Bonnet and Lamarck use the same cast of characters—God, nature, the chain—but they are given different roles to play and the tone is not the same. Bonnet's God presents Himself as the God of Creation and teleology; He is an object of wonderment. Nature speaks a divine language—she transmits divine revelation by refracting it. The chain extends beyond living beings up to the "celestial hierarchies" (part 1, chapter 12), up to the angels. Theological structure penetrates deeply into scientific observation. Lamarck, on the other hand, brings in God as a rational principle. He seems to respect religious convictions. But does he share them? Bonnet invokes God as "the adorable hand." Lamarck questions and builds; if impulsiveness takes over, it is that of logic. His ardor is knowledge. Bonnet marvels, and religious emotion pervades his writing. Lamarck is in the throes of scientific exaltation.

The Lamarckian series begins with a new intention. By shedding light on the genesis of his project, the analysis at the beginning of this chapter also sheds light on the project as a whole. The latter is rigorously scientific—to do away with species as a rigid entity and as a support of classification that poses as science; and to generate the natural order, which is the only authentic object of knowledge. Lamarck's series is the key to this project; or rather, it would have been the key if the coherence it had postulated had not been altered in its internal structure. But our critique has uncovered some vacillations. One searches in vain for an intelligible necessity behind the organization of each mass and the process of degradation that allegedly connects these masses. We saw that the series was neither shown nor proved. This led us to a third stage—an examination of the idea of series in philosophical-scientific thought—in order to discover its status. We very summarily reduced the types of series (and concepts similar to it) to three examples spread out in time: Aristotelian philosophy, Leibnizian philosophy, and the ideological-scientific vision of Charles Bonnet. A more detailed analysis would have followed the concept of series through the works of nearly all post-Leibnizian savants. But the Lamarckian series could be reduced to none of these earlier models, and in contrast to them its own originality became more apparent. This series was first of all chosen—indeed it could scarcely not have been chosen—because the natural order is a succession. Unfortunately, by the same stroke, all the ambiguities associated with "nature" returned through the back door. On the one hand, the series provided a basis for the concept of evolution by suggesting an increasingly complex order of productions; on the other, it took a stand against creationism. Nature would thus be saddled with the vocabulary of fixism that it was no longer desirable to associate with God. Transformism was not introduced by the inversion of the series. This inversion was a necessary but not a sufficient condition—to read from bottom to top what was intended to be read from top to bottom scarcely constitutes a serious justification for the transition to a chronological mode. It is in fact the concept of "circumstance" that makes the transition from the purely logical ordering to the order of production through the existential concept of change.

The idea of series thus goes through three decisive stages:

1. the establishment of a natural order;
2. the discovery of the succession of productions and of their perturbations (a linear model overloaded with branches); and
3. descent, which has not yet been discussed.

Change is introduced by the facts themselves, since the methods of breeders and the experiments prepared by nature actually show species or varieties undergoing modifications before our very eyes. Lamarck did not go so far as to perform his own experiments. He was, if I dare say so, an observer of the experiments of others (breeders) and of nature. He did not raise pigeons like Darwin, but he did have plants in the ground and "collections" in the "cabinets of natural history." Lamarck knew, however, that a phenomenon must not only be observed but also explained. It is here that the laws of use and disuse, and of the inheritance of acquired characters, enter the picture.[44] The introduction of transformism could be demonstrated only if transformations were founded on truly scientific concepts. The question was therefore the following: Is there truly a theory that links collections of facts to the ideas suggested by these collections? This is the theoretical void that Darwin evidently tried to fill during the twenty-one long years between the return of the *Beagle* and the publication of the *Origin of Species.* In Lamarck's work, what is the theoretical level of the following chain: nature-series-circumstances-anomalies-acquired variations-hereditary transmission of acquired characters? This question can be answered only after examining the last two terms.

From the point we have reached, we can only say with certainty that the series is a hybrid concept, and this influences the way in which Lamarck uses it. It stands at the crossroads of a philosophicotheological heritage and a scientific research project. We can state that thanks to the series, Lamarck did indeed devastate the creationist position. In this, he was no precursor. Instead, it was rather his colleagues who were backward and fainthearted; they either did not see the scientific interest of the operation or did not want to see it. Lamarck did precisely what needed to be done at the time. While adopting the ideas of the entire eighteenth century about living beings—scattered ideas that announced evolution more than they formulated it and in any case did not turn them into a system—Lamarck ordered innumerable facts into his "philosophy," which is in fact ideological-theoretical. Out of it he brought this important whole: transformism, which overturned the organic world view. But in so doing, he did not eliminate the ideological notions that pertained to the religious in nature, the old idea of a scale of beings. He even retained the ancient order of this chain of concepts, even as he was working on each concept from the inside. Thus the God of whom he spoke was neither the God of Christian revelation, nor the God of Rousseau, nor probably even the

God of Robespierre. It was rather the God of "the philosophers and savants," whom Pascal had rejected and Bergson would analyze as an "abstract concept."[45] Nature, whose mysterious power François Jacob mentions,[46] was not demystified, as is obvious in the serial plan of increasing perfection and the idea of natural harmony. Yet it proved to be a network of rigorous causalities as it unfolded. It did not operate within the chain of arguments, but only at the beginning of the argument.

The analysis undertaken in the preceding considerations of nature and series in Lamarck aimed to determine their precise status. We have distinguished aspects that belong to the scientific procedure of the savant from those that betray an ideological "collage" at the point where scientific features do not have the means to emerge. We have yet to base this distinction on a general epistemological view and a theory of ideologies. The work of Georges Canguilhem and Louis Althusser, on different planes, will help to sketch the procedure.

In one of his latest books, Georges Canguilhem has analyzed scientific ideology *(idéologie scientifique)* and the ideology of the scientific *(idéologie du scientifique)* from an epistemological perspective.[47] He suggests that these two closely related expressions should not be confused. Scientific ideologies, which are produced "rather" by philosophers (the "rather" is important), are "discourses with scientific pretensions by people who are, with respect to the matter at hand, only presumptive or presumptuous scientists."[48] The ideologies *of* the scientific are "ideologies that scientists generate in the discourses they make to thematize their methods of research and encounter with the object of study, in the speeches they give on the place science occupies within culture, and in relation to other forms of culture."[49] These ideologies are philosophical in nature. Spencerian evolution is an example of scientific ideology. Buffon's "organic molecule" and Bonnet's "scale of beings" also pertain to scientific ideologies and are given as examples of them. On the other hand, the concepts of "nature" and "experience/experiment" *(expérience)* in the eighteenth century are examples of the ideology of the scientific. Can this set of definitions and illustrations shed light on Lamarck's *zoological philosophy?* It seems to me that "nature" is for the most part a concept drawn from the ideology of the scientific. Indeed, in Lamarck, it thematizes a method of research that leads to the "natural" order, as opposed to classification as "art." It also thematizes the relation with the object that is apprehension of the organization of living things. It locates science within the realm of culture by establishing the relation of biology to physics, chemistry, geology, and hydrogeology on the inorganic side; and its relation to the system of man's positive beliefs on the anthropological side. These relations are the cultural structures of nature, which begins with the mineral and culminates in man.

As the implementation of the plan of nature, the series cannot be dissociated from it. The Lamarckian series is therefore integrated into this ideology of science.[50] The concept of series sketches out nature herself more precisely than the concept of "nature." Through the series of living beings, and of animals in particular, nature appears in her totality, including the environments *(milieux)* that act through "circumstances." And these environments extend to systems of stars, if not to legions of angels. This last point raises in a light-hearted way the problem of the relation of the Lamarckian series to Bonnet's "scale of beings," which Canguilhem treats as a "concept of scientific ideology." If Bonnet's "scale of beings" is the ancestor of the Lamarckian series, how ought we consider the latter? Its relation to nature and its function in scientific practice place it in the ideology of the scientific, but its relation to Bonnet's scale of beings seems to draw it toward scientific ideology! The problem is twofold: Is it indeed true that the scale of beings and the series are in continuity? And how can the scale of beings be a concept of scientific ideology if it sprang from the head of a naturalist? The answer to the first question lies perhaps in the answer to the second.

In Henri Daudin's classic study, Bonnet's scale stands chronologically before the Lamarckian series. Daudin knew very well the important role Leibniz played in Bonnet's thought. It is evident that this "scale of beings" is not at all the discourse of a scientist, but of a philosopher who inhabits the scientist, be it himself or another. As we have seen, the concept of "scale of being" is theological-metaphysical; it borrows part of its hierarchy from the sciences; the remainder is a "presumptuous" rather than "presumptive" extrapolation. In contrast, the series is the patient product of painstaking scientific work, even if imperfect and unsatisfactory. Its ideological element is internal to this scientific practice. Canguilhem's fruitful distinction makes possible the dissociation of what Daudin had only aligned on a temporal axis. Two concepts that apparently stood on the same level reveal two radically different functions when examined in light of Canguilhem's distinction.

To explain satisfactorily this distinction between two ideologies intimately connected to science, I have appealed to the fact that the scientist is sometimes possessed by a philosopher, or at least by a philosophy. But a theory of ideologies should not be made to depend on a psychological remark—one must grasp the theoretical operation of a thought, an internal structure. Louis Althusser, the theoretician of ideologies and their relation to science through the scientist, brings into play the "characters" of his "little theatre": their names are SPS and COW;[51] their stage is scientific practice. Like those Russian dolls that are fitted one within the other, they act inside each other, within the one that contains them all—scientific procedure. Althusser's machinery is set into motion by a materialist thesis that is obvious from the outset. This thesis is simulataneously a speculation and a combat position. The point is to show that at the very core of his activity, the scientist is imbued with the idealism of the ruling class. He can only become free by leaning on the materialism of the rising classes that struggle against these ruling classes and ideologies. These fundamental theses immediately reveal both their distance from and relation to Canguilhem. Indeed the latter explicitly claims to have introduced the concept of scientific ideology into his teaching in 1967–1968 "under the influence of the work of Michel Foucault and Althusser,"[52] and does Althusser the honor of criticizing his ideas and those of Dominique Lecourt.[53] The dependence in which the production of knowledge stands in relation to the social activity (class struggle) that provokes the bending of "the elaborated purification of verification norms" crushes the epistemological realm under the weight of dialectical materialism. Truth and falsehood are displaced, if not replaced, by "conformity to one program," as Canguilhem has noted with severity.[54] In the perspective of the Althusser school, Lamarck, like any other scientist, has a spontaneous philosophy within his scientific practice: "By SPS, we mean not the set of ideas that scientists have about the world, (i.e., their world view), but only their ideas, whether conscious or not, about scientific activity and science."[55] This SPS is not monolithic, but divided by a contradiction: two of its elements struggle with each other. The first element is intrascientific; it concerns (1) the real existence of an object of scientific knowledge, (2) the objectivity of scientific knowledge of the object, (3) "the correctness and effectiveness" of the scientific method.[56] It looks as if we find here one aspect of the "ideology of the scientific" analyzed earlier, but accompanied here by the adjective "spontaneous." This element is materialist in comparison with a philosophy that would introduce questions of principle from the theory of knowledge. This first element of the SPS supports all we have considered truly scientific in Lamarck's studies of nature and series. If we concede that another element, idealist and extrascientific, which pertains to scientific practice but comes from elsewhere, introduces "values" or "demands" drawn from "religious, spiritual, or idealist philosophies," then we introduce the dialectical contradiction. Now it seems that the recourse to the Supreme Author, the mysterious character of "nature," the appearance of the series as a design of this nature, herself delegate of the Supreme Being, are all extraneous elements drawn simultaneously from theology and law. To understand the complex structure of this presumed SPS for Lamarck, it is of course necessary to turn to the philosophy of the Enlightenment, the state of the sciences, the political regimes the scientist experienced. We have done this. I merely want to recall one point, which warrants emphasis. The second, idealist, element of this hypothesis is dominant since the world of the scientists reflects society. Althusser states the impossibility of reversing this relation of dominant to dominated without appealing to an external factor that tilts the scales in the other direction. According to him, at present this factor is dialectical materialism as an ideological and political position. In the eighteenth century and at the beginning of the nineteenth century, this factor was the philosophy of the Enlightenment.[57] It was beset, moreover, by the contradiction of materialism on the one hand, and deistic, judicial, or other forms of idealism on the other, so that Lamarck would have both been aided and hindered by the heritage of the eighteenth century. But beginning with the Empire and under the Restoration, the religious apparatus acted as an auxiliary to the ruling politics in a manner such that even nonconfessional deism, even ethical-juridism, in short, even idealism, could serve as a tool to fight the censure exercised against biological evolution and generally against the annoying presence of the divinity in the scientific realm. Having said this, it is clear that Lamarck trusted the weapons of knowledge, as he noted explicitly in his "system of positive knowledge." His support for the Revolution was superimposed upon the academic and nonrevolutionary belief that the world could be changed by proving scientific truth.[58]

We shall not push any further the test of an analysis inspired by Althusser; to make it complete, it would be necessary to extend it to the level of "philosophy" and of "world view." Yet it must be admitted that this analysis clarifies many a point, especially the presence of a philosophy within the scientific procedure itself, and not only in the discourse that the scientists make to thematize this procedure.

Insofar as we hold fast to the idea that a science is never free of elements foreign to the scientific enterprise, we are constantly searching for the very meaning of the term "science." It is as a consequence of having overlooked this equivocation that some have fallen victim to the myth of the precursor with regard to Lamarck. One need only restore to him the complexity of his practice and the amalgam it represents to bring to light the indubitable fact that his heart beats with his own era.

4 *The Transmission of Acquired Characters*

The concepts of "nature," "series," and "circumstance" are susceptible to partial conceptualization even though the presence of a nearly irreducible ideological element makes such an attempt very difficult. This is not the case for the expression "inheritance of acquired characters," which is later "collage" based on what Lamarck preferred to call "the preservation of acquisitions" by "generation" for "new individuals," that is, the descendants of those that the circumstances modified. He also referred to the transmission of "acquired changes." These terminological differences are significant; they emphasize the differences between the various realms of the possible at the beginning of the nineteenth century, the end of the nineteenth century, and the beginning of the twentieth century. In 1809 "heredity" was not a scientific concept. It became one first in medicine.[1] "Character" was a term used by Darwin, made more precise much later by Mendel and emphasized by genetics. As for the concept of "acquisition," the only one that seems reasonable to use in reference to Lamarck, it was never elaborated with precision, in spite of appearances to the contrary. As we shall see, its content underwent a metamorphosis at every stage of the history of biology.

Far worse is the fact that biologists after Lamarck abusively applied their own terminology to him and thus completely obscured his own statements and significance. The general neglect into which Lamarck first fell (in spite of a few isolated supporters) and his resurgence in France and the United States in the late nineteenth century following the diffusion of Darwin's theory of evolution do not make the task any simpler. Far from noting a continuous linear development of questions and answers, one sees conflicts, breaks, returns, and pseudoreturns. (For just as the concept of precursor has no theoretical content, so the concept of the resurrection of a theory fifty years after it first appeared is a pseudoidea.) The so-called "theory of the inheritance of acquired characters" is therefore a misnomer. It would certainly be useful first of all to denounce its equivocations.

The approaches to the problem are multiple: The level of visible forms would be followed by the organic level; the latter by the cellular, genetic, and molecular levels. To the consideration of individuals would then be added that of populations, which would give importance to the quantification already under way in Mendelian genetics. Finally, the debate of the acquired and the innate would become entangled in the ideological net of Stalin's policies at the time of the all-too-famous "Lysenko affair" in 1948. Currently, the racist, antiracist, democratic, and antidemocratic implications of the debate are further altering the terms of the argument.

The requisite procedure involves first of all a return to the Lamarckian documents and to Lamarck's "environment."[2] The key idea that emerges from a study of the texts in their historical context is the following: For Lamarck, the "inheritance of acquired characters"—an expression we shall tolerate in quotation marks—was to be sure a necessary support of his theory, but it was not a new thesis. No one either defended or contested it. It was not even an issue; it simply went without saying. After Weismann took another look at Darwin, the problem of heredity became a real question and the quasi-dogma of acquired characters came under attack. Then neo-Lamarckism entered the arena to defend the transmission of acquired characters and brought into the foreground of the debate an issue that originally had never been in question. And at this point, it obscured the true debate. It was nevertheless clear that if the transmission issue were to be brought into question, the whole of Lamarckism would be at risk. The serious mistake is the failure to see that although this question was neo-Lamarckian, it was not Lamarck's. We thus return once again to the notion that the questions Lamarck was asking were not prophetic, but fit squarely into their own era. The dislocation does not consist in considering the transmission of acquired characters, but in the necessity of demonstrating it from this time on. The emergence of this necessity was nevertheless the cause of the displacement of the heart of the debate to heredity. And it is often forgotten that at the same time, between 1800 and 1830, the effects of use and disuse upon organisms and the modifications of species as a result of "circumstances" constitute the sore point at which transformism is at stake.

The *Philosophie zoologique* is the essential text. We are not overlooking the fact that the principles discussed in this chapter were already present and had been elaborated in the *Recherches sur l'organisation des corps vivants* (1801), the *Système des animaux sans vertèbres* (1802), and the opening lectures of Lamarck's courses.[3] 1809 was not the first year in which the transmission of acquired characters was mentioned. It was, however, the first time it was integrated into a theoretical synthesis. The philosophy of a science represented for Lamarck the structure of its completely unified principles, the possibility of a general overview of its investigations. The *Philosophie zoologique* stood at the center of his thought, and also at the center of his life as a researcher; it followed many partial accomplishments and preceded his year-by-year work on the majestic *Histoire naturelle des animaux sans vertèbres*, after which he would publish his last work, an attempt to construct a positive anthropology: "I could prove that it is neither the shape of the body nor of its parts that gave rise to the habits and to the way of life of animals. On the contrary, it was the habits, the way of living and all the influencing circumstances, that, with time, shaped the body and parts of animals."[4] Thus that which in 1809 led to the law of use was presented in 1815 as the most important consideration. The transmission of acquired characters was not even mentioned. But this principle was formulated in Year X (1802) and referred back to the primacy of the functions over the organs. The *Discours de l'an XI* (1803) picked up both themes. The *Avertissement* of the *Philosophie zoologique* states that "the steady use of an organ contributes to its development, fortifies and even increases it, whereas a habitual lack of use in an organ hinders its development, deteriorates it, gradually reduces it and ends by making it disappear if the disuse extends to long periods of time, etc."[5] Significantly, the transmission of acquired characters is not mentioned here. It is in chapter 7, on the "influence of circumstances," that the two laws of transformism appear:

First Law
In every animal that has not reached the term of its development, the more frequent and steady use of any given organ gradually strengthens this organ, develops it, increases its size, and gives it a power proportional to the duration of this use; whereas the constant failure to use such an organ imperceptibly weakens it, deteriorates it, progressively decreases its faculties and finally makes it disappear.

Second Law
All that nature has caused individuals to gain or lose by the influence of the circumstances to which their race has long been exposed, and consequently by the influence of predominant use of a given organ or the constant disuse of a given part, nature preserves by generation for the new individuals that issue from them, provided that the acquired changes are common to both sexes or to those that produced these new individuals.[6]

The second law ensures both continuity and the transformation of species, which as we have seen has only a relative value within the "natural order," but remains an unquestionable unity in this order as well as in "art." Transmitted modifications cause an ever more complex diversification and therefore transitions between specific groups that earlier had been clearly separated. The concept of a distinct group nevertheless remains valid, even if the differentiating criteria are modified or weakened. Species are the primary reference point of transformations. With the *Philosophie zoologique*, the connection between concepts emerges very clearly. "This admirable work is the first reasoned exposition of the genealogical doctrine that is taken to its logical conclusion," wrote Haeckel in the preface to the 1907 edition. The first six chapters[7] offer the following conceptual chain: *opposition* of the *artificial systems* ("art") to the *natural order*,[8] from which *nature* and the limits of the concept of *species* are drawn;[9] the division between the *organic* and the *inorganic*; *series* as the design of organized being; *circumstances* as obstacles to the production of a serial plan; stimulating *needs*, which give rise to *actions* that modify the organs and lead to the diversity of species by transmission from one generation to the next. So it is that we reach the transmission of acquired characters.

There is one curious lacuna, however—the concept of adaptation is missing. The word is nowhere to be found in Lamarck. If the term is absent, is the phenomenon covered by another concept? Famous biologists have referred to Lamarckian adaptations and even cited Lamarck's own examples (the mole, the aspalax, the giraffe, etc.) as instances of adaptation. What should one say about this concept, or about this phenomenon, conceptualized or not? A special study[10] will not be superfluous in order to see more clearly into a question that Camille Limoges's excellent book raised.[11]

The law of the transmission of the acquired in chapter 7 of the *Philosophie zoologique* is integrated into a synthesis. It is tied to use and disuse by the concept of circumstances, and by the modifications that issue from them. From here they are linked with the series, where they play a perturbing role, by producing varieties that the design had not foreseen. The modifications depend on physical factors (climate, water conditions, food, etc.), but these factors also create needs, actions, habits, which imply relations between the physical and the moral. The *Discours préliminaire* quotes Cabanis frequently, but Lamarck did not think the latter had paid "enough attention to the influence of the moral on the physical."[12]

Cuvier's eulogy, as disseminated by the Edinburgh school, certainly is at the root of a simplistic interpretation of Lamarck. Desire was substituted for needs, so that the former appeared to be the direct cause of organic modification. In a well-known passage, Darwin wrote to Hooker,[13] "Heaven forfend me from Lamarck nonsense, of a 'tendency to progression,' from 'adaptations from the slow willing[14] of animals.'" Thus Darwin himself referred to Lamarckian adaptation. Yet Lamarck not only referred to *need* where one imputes to him an appeal to desire and will; he also carefully distinguished the "sentiment" of irritability[15] or even the "eminent faculties" proper to man from the manifestations of inferior animals. Cannon[16] is right to denounce the misdeeds of Cuvier's eulogy. With all due respect to Richard Burkhardt, who recently published a remarkable book on Lamarck, Cuvier's hostility is not a legend. It is therefore appropriate to respect the "primacy and the universality of need."

Some people have been surprised to see the two laws from the *Philosophie zoologique* reappear several years later in a set of four laws within the *Histoire naturelle des animaux sans vertèbres*. The first law is new: It formulates the tendency of life to increase "the volume of the body it possesses and to extend the dimensions of its parts up to a limit set by itself." The second law concerns the production of a new organ following a new need. In my opinion, this is the most spectacular case of the effect of "use" on the organs, which is the subject of the third law. One should note here that this third law succinctly rephrases the first law of 1809, and that the birth of an organ has become the subject of the second law, a topic mentioned in the text of 1809 but not in the laws. The fourth law takes up the transmission of acquisitions. Its obviousness is affirmed on the basis of innumerable observations, although several exceptions are conceded. There is no contradiction between the texts of 1809 and of 1815—neither the place nor the function of the transmission of acquired characters has changed within the chain of reasoning.[17] But Cuvier was not interested. He would save his sarcasm for the law of use and disuse of organs; this is the one he deemed important to demolish: "It is by dint of wanting to swim that membranes come to the feet of aquatic birds; by dint of wanting to go into the water, by dint of not wanting to get wet that the legs of shore birds get longer; by dint of wanting to fly that the arms of all turn into wings, and that the hair and scales develop into feathers. And let no one think that we are adding anything or leaving anything out—we are using the author's very own words."[18]

Let there be no mistake; there is more here than the light skirmish of an academician. Within this urge to heap ridicule upon Lamarck's discourse lie the rumblings of the threatened ideology of fixist creationism. The text under attack is the following:

> The bird whose needs draw it toward the water to find the prey that gives it life spreads the fingers of its toes when it wants to hit the water and move at its surface. The skin that binds the base of the fingers contracts the habit of stretching through these constantly repeated spreadings of the fingers. Thus with time, the large membranes that unite the fingers of ducks, geese, and so forth, were formed such as we now see them. . . .[19]

The texts can be compared and evaluated. It is noteworthy that Cuvier attacked only the law of modifications by use and disuse, a law that Lamarck defended by *reductio ad absurdum* and by appeal to the facts: (1) If the principle were false, nature would have created "as many forms as required by the diversity of circumstances in which they are forced to live," and it would have been necessary for "these forms never to vary."[20] (2) On the other hand, what might be the reason behind the existence of "race horses" and "draft horses," of "basset hounds" with "bow legs" and "greyhounds so adept at running," of "chickens without tails" and "peacock pigeons?"[21] The argument from breeding, which is so important for both Lamarck and Darwin, is already beginning to emerge. "What nature does with much time, we do everyday by suddenly changing the circumstances in which a plant and all the individuals of its species were found."[22] Among the examples brought to bear on the argument are cultivated wheat, cabbages, lettuce, and once again chickens and pigeons, caged birds and domestic dogs. Lamarck's transformism rested only on what has been called the law of adaptation; the second law (the transmission of acquired characters) was implied in the first—it went without saying even though it was said. So it was that after having based everything on the divine plan manifested in the series, Lamarck buried the latter under a diversity of forms produced by the diversity of circumstances in a nature capricious enough to stand in opposition to herself. But the obstacle had become an agent. The anomaly existed only with respect to the serial norm; it was perceptible only in the decrease and deterioration of organs: The mole loses its eyes. The aspalax, which "lives underground like the mole," loses them like it. The proteus, "which inhabits deep dark caverns," becomes blind. The adult whale has lost its teeth. (Their existence had remained unknown until Geoffroy Saint-Hilaire discovered them in the foetus). Yes, all of these cases could be termed anomalies. But what were the palmations or the long legs of shore or fishing birds? *Felix culpa!* Beneficent anomalies. Lamarck's followers would not hesitate to call these—rightly or wrongly—"adaptations."

This way of arguing, this series of observations leads into a closed realm where two perspectives oppose each other: The first, fixist creationism, is ideological; the second, transformism, is predominantly theoretical. "It is a *conclusion conceded to this very day* that in creating the animals, nature (or its Author) foresaw all the possible types of circumstances in which they could have lived, and gave each species a constant organization, as well as a determinate form invariable in its parts, which force each species to live in the locations and climates where it is found and to preserve there its known habits."[23] The adjectives "constant," "determinate," "invariable" express the double fixity of external factors and organisms, the correlation between the two orders of facts. But the same correlation between organization and environment might suggest another result: "My own particular conclusion is that, in producing successively all animal species and in beginning with the most imperfect or simple in order to end with the most perfect, nature has gradually complicated their organization. Since these animals generally spread themselves out through all the inhabitable parts of the globe, every species received from the influence of the circumstances in which it found itself the habits that we associate with it and the modifications of its parts that observation shows us in it."[24] Here the concept of succession is opposed to that of constancy (first conclusion); it is linked to the concept of "modification." Both conclusions bring natural causes into play, but the first is based on creationism, the second on the concept of transformation—hence the absence in the latter of the parenthetic reference to the Divine Author, Who is quietly removed from the realm of science. "Can there be in natural history a consideration more important and to which more attention ought to be devoted than the one I just presented?" asked Lamarck.[25] The consideration mentioned here does not pertain to the mode of formation of organisms. (The transmission of the acquired is understood.) This is the confirmation of the thesis we presented at the beginning: Lamarck was not behaving like a precursor, but simply like a penetrating mind who interpreted and synthesized the possibilities of his era with the means at his disposal. This is the reason why he did not raise the question of the transmission of the acquired—it was not a question. That some of Lamarck's colleagues, as good practitioners and not so good theoreticians, were unable to exploit the possibilities of their present does not make Lamarck a precursor. Lamarck was in tune with his own times. Hence his perceptiveness, but also his lapses and lacunae. To be in tune with one's times requires a rare perceptiveness and much boldness.

Even a mere outline of the history of science confirms the remarks of François Jacob in his history of heredity, the statement of R. Burkhardt in his recent book, and Weismann's conclusions in the nineteenth century—the inheritance of acquired characters is a very old story that is not called into question. Until Weismann, "the possibility of inheriting acquired characters was never seriously doubted." The Book of Genesis recounts the story of a certain patriarch who pealed poplar, almond, and maple branches so that his flocks would yield striped, speckled, and spotted young after breeding in full view of the striped branches. So it was that he became rich at his father-in-law's expense.[26] Perhaps Adam's sin is the unforgettable archetype for the transmission of an acquired character. The Egyptians and the Greeks thought that children carried within them the experiences of their parents. After Descartes, Malebranche claimed as much in his *Recherche de la vérité* in the seventeenth century. The transmission of the acquired certainly slowed down the analysis of living beings and reproduction. But it should be noted that at the beginning of the nineteenth century, it filled a void that no scientific datum could fill. Cell theory still had its problems; Auguste Comte had condemned the early research in the field. It was only in the second half of the century that the work of Dujardin, Purkinje, Schwann, and others became established in all of its breadth. "Generation," the term used by Lamarck, was still imbued with the idea of creation; "reproduction" was certainly a term in use since Réaumur and Buffon, and with it the natural tended to be substituted for the supernatural. But in 1881 Weismann could still write that "nothing is known of heredity." The past brought to Lamarck what was a dogma for some, but self-evident for him. He did not ask himself the question of heredity, whereas Darwin did, and answered with the theory of "pangenesis."[27] Gemmules are transmitted by the parents to their offspring; they come from every cell and join to form the sexual elements. Gemmules transmit the modifications that have affected the cells; they fuse in the course of development. This theory had an explanatory value for acquired characters as well as spontaneous variations. Although Weismann did not accept this Darwinian hypothesis, he nevertheless thought of it as a necessary step.

It was Weismann who turned the problem on its head: He declared that the transmission of acquired characters is impossible. He thereby opposed not Lamarck's second law, as Cuvier had done, but rather the first. He thus made the inheritance of acquired characters the central theme of Lamarckian theory, which it had never been at the beginning of the nineteenth century, but suddenly became so as soon as it came under fire. It is useless to recall the total separation of the germ cells and somatic buds. The somatic buds that make up individuals are only "little isolated plants" that arise "some distance apart" from the "long tracing rhizome," which is eternal. Every variation is germinal; the soma has no evolutionary value. This view spared nothing of Lamarck, whereas Darwin, thanks to natural selection, could explain the disappearance of nefarious organs and the proliferation of useful variations. He allowed acquired characters, but these were not essential. On the contrary, as Weismann noted, "If acquired characters cannot be transmitted, then the Lamarckian theory completely collapses, and we must entirely abandon the principle by which alone Lamarck sought to explain the transformation of species—a principle of which the application has been greatly restricted by Darwin in the discovery of natural selection, but which was still to a large extent retained by him."[28] In other words, natural selection had replaced use as the dominant concept. The Weismann-Lamarck debate, if we may call it such, did not discuss the fact of transformations, but rather their factors. It was a struggle between two factors, a struggle that turned on the germinal-somatic opposition. During the Cuvier era, the goal was to strike at the transformations themselves by caricaturing the use factor. Around 1830 the ideological debate between transformism and creationism stood in the foreground. But by the later nineteenth century, two forms of evolutionism were doing battle. Weismann's conclusions were solidly founded on cell theory, on the work of Nägeli and others. But Weismann was wrong to give credence to arguments that confused mutilation and acquisition and to infer the nontransmission of the second from the nontransmission to the first. From Weismann on, the defense or criticism of Darwin corresponds to the criticism or defense of acquired characters. One need only examine the neo-Lamarckians of the late nineteenth century to be convinced that the defense of Lamarck hinges on the defense of acquired characters. It is then that the uncertainties in the meaning of the term "acquired" begin to emerge and that changes in its meaning appear.

Far from being discouraged by Weismann's attacks, the neo-Lamarckians stood very firm. In the United States, the great paleontologist Cope, who may be considered the leader of New World neo-Lamarckism, searched fossils for evidence of evolution. He defended the influence of habits and, consequently, acquired characters. In France, the movement was represented by Edmond Perrier, Gaston Bonnier, Félix Le Dantec, and Alfred Giard.[29] The latter, according to Burkhardt, did not know Lamarck very well. I do not think this doubt is justified. Some of the introductory lectures for his course on the evolution of organized beings (1889, 1898) focus on the primary Lamarckian factors of evolution. Thus Lamarckism is defined as the study of the primary factors of light, temperature, climate, food, nature of the waters, and so forth—in other words, those factors that provoke the needs, habits, and actions (and therefore the acquisitions) and presuppose that they are hereditary. The last point was the one that required proof, since Weismann wanted to exclude it from science. "If this principle of Lamarck is not exact and cannot be demonstrated, it is clear that the role of these primary factors is significantly decreased."[30] These factors are the circumstances of the environment, tied to the use and disuse of organs. The secondary factors are Darwinian: natural and sexual selection and so forth.

Giard alludes to a talk that Weismann delivered on 20 September 1888 in Cologne in which he claimed, "I think I can state today that the material existence of a transmission of acquired characters cannot be demonstrated and that there exist no direct proofs of Lamarck's principle."[31] Here the meaning of the word "acquired" rests on the distinction between "somatogenic and blastogenic modifications." Only the latter reach the reproducing elements. It should be noted that the problem here takes into account new cellular data that make possible a distinction between the soma and the germ plasm. But on the other hand, the acquired is defined as an influence sufficiently pervasive to affect the germ plasm. This view is of course based on the refusal to concede absolute independence to either type of cell. Lamarck would simply have said that the pressure of circumstances directs the "fluids" and "forces" toward the "part" of the body with the new need.

The difference between the concepts of acquired character at the beginning and at the end of the nineteenth century is now evident. With neo-Lamarckism, reproductive cells are at issue; they cannot be at the time of Lamarck. In both cases, the influence can and must be great enough to become hereditary. Thus, for Giard, acquisition has a different meaning than for Weismann. It is surprising, however, that in his Freiburg lecture of 21 June 1883, Weismann did not take into account the research of Brown-Sequard. In a lecture given on 13 March 1882, Brown-Sequard had included in the category of transmissible acquisitions accidental lesions caused by the section of some nerves.

Thereafter, the concept of acquisition went through several

levels of transformation, and the Lamarckian meaning was disfigured. Giard seems to have overlooked the fact that many scientists used Lamarck as a weapon against Darwinist "materialism." Giard himself turned Lamarck into a defender of democratic ideas; he quoted Picavet, Landrieu,[32] old articles quoted by Picavet or some author who in 1847 presented Lamarck as a defender of freedom, and texts pertaining to the revolutions of 1830 and 1848. In the preface to his *Discours*,[33] he noted that Lamarck's ideas glowed anew during every period in which freedom had to be defended. In this case, the acquisition of acquired characters was linked to a democratic ideology. It was thought advantageous to present Cuvier as a defender of the "ruling classes" and religion. (One must bear in mind the strong laical [anticlerical] movement at the turn of the century.) Thus the transmission of acquired characters was openly and deliberately linked with a political ideology favorable to the sciences, so that the two interests were conjoined. These were the years of the great Lamarck biographies by Alphaeus Packard in the United States and Marcel Landrieu in France.

Today, the neo-Lamarckian current still has its representatives. Some, like Father Grassé, nevertheless give Darwin his due;[34] others, like Wintrebert, are intransigent zealots for neo-Lamarckism. The latter claimed in 1962 that "the inheritance of acquired characters, not yet demonstrated by experimentalists, is the point of contention."[35] If one wants to make headway in this question, a new meaning of "acquired character" is necessary: "The living does not pass on what it receives, but what it does."[36] He adds, "I have no doubt that it would be possible to demonstrate not only the inheritance of acquired characters but also the chemical mechanism of its acquisition, provided one placed living beings in nature as Michurin and his disciples did in Russia, that is, by first giving the organisms themselves a chance to effect an adaptation; then, once they have acquired it, by using biochemistry to uncover the presence of an adaptive hormone and cytology to show the appearance of a new gene."[37]

One should note, after thirty years, the appeal of Michurin, whom Lysenko had drawn on. Wintrebert judges the latter severely, but without going too far in his criticisms.[38] He is not trying to be soft on Stalinist ideology, but probably wishes not to attack too harshly a theory based on the influence of the environment. The environment primes the gene; but the environment also provokes the creation of new genes. "Function creates form," as Geoffroy Saint-Hilaire used to put it, and in order to create new forms, new genes must be created.

This function creates the deoxyribonucleic hormone chemically, which combines with the nucleoprotein of the species by forming a gene, which in turn mutates the defective protoplasm by adapting it in time and space. Everything happens on the functional plane and it is the living that adapts. The new form is only the morphological translation of the newly created chemical entity. *The hormone is the chemical test; the gene is the cytological sign of its incorporation into heredity; the form, the visible manifestation of the created protoplasmic mutation.* Lamarck could only consider the form, and indeed it is completely unfair to hold this against him.[39]

Such is the essence of this "chemical Lamarckism," which I shall refrain from evaluating. "The gene is acquired and transmitted," Wintrebert dares to claim. Father Grassé assesses this contemporary neo-Lamarckism with a judgment that is both conscious of its success and reserved about its conclusions: "Wintrebert lacks the support of experiment."[40] That is quite a lot. By refusing to evaluate the basis of the theory, he praises it for giving to "need" the role that a certain brand of materialism attributes to chance. Grassé is not an unconditional supporter of neo-Lamarckism, but he firmly rules out chance as having biological or objective significance; he attaches much importance to the reaction of the living being to its environment. He is very open to all the research that casts doubts on the theory of innate variations, yet he welcomes selection and the legacy of Darwin and Weismann, but without adopting the latter's outdated and dogmatic conclusions. Above all, Father Grassé recalls a crucial element, which the discussion of heredity has obscured and to which neo-Darwinians paid no attention: the whole of paleontological data. The role of fossil shells in Lamarck's thought is well known; we alluded to the importance of the neo-Lamarckian paleontologist Cope. There is an authentically Lamarckian connection with paleontology, and it begins with conchology. Fossil shell evidence made possible a counterargument to Cuvier's theory of catastrophism and a confirmation of evolution. "Every explanation of evolution that loses sight of paleontological data is only a theory in which the imaginary plays a preponderant role."[41]

The transmission of acquired characters became more complex as the theory of evolution became more synthetic. It drew not only on paleontology, embryology, and cytology but also on genetics, molecular biology, and so forth. All of these sciences were possible approaches to it. Weismann, who marked a crucial turning point in the nineteenth century, did not for that reason become a museum piece, as one can easily see by surveying the present state of evolutionary biology,[42] but heredity was translated into the languages of the new sciences. From another point of view, the relation of DNA to RNA, to the amino acids, to proteins, is analogous to the relation of the germ plasm to the soma. In Wintrebert and others, the old quarrel is reemerging, about which Lucien Cuénot remarked in 1925 that it is always reborn from its cinders when one thinks it dead. "I am opposed to acquired characters," he wrote, but added that "if one experiment could be established in their favor, the results would be considerable."

The death of acquired characters is a certainty for Jacques Monod. I would have said as much for François Jacob, had I not read his article on teratocinoma and cellular differentiation,[43] which summarizes and adapts his lecture before the Royal Society in 1978. But I would not have had the idea of a Lamarckism in Jacob, and I admit that I have not adopted it, even though I have had the privilege of becoming acquainted with the reaction of Marek Glogoczowski,[44] who considers the developmental schema of the normal organism to be "strictly similar" to that of Lamarck on the inheritance of acquired characters. This bold statement is based on the fact that the cells all contain the same DNA, and yet are differentiated by interaction. It thus appears that they acquire new structural elements. Glogoczowski's Lamarckism is in my opinion more obvious than that of François Jacob's and is especially effective in bringing out the vitality of the question! Much research seems to cast doubt on categorical affirmations like those of Peter Medawar: "The translation of genetic into structural information is irreversible."[45] At the very time Jacques Monod was writing that information is never transferred from RNA to DNA,[46] the work of Temin, Baltimore, Spiegelman, and others was undermining his assertion. "The discovery of enzymes capable of using viral RNA as a matrix for the synthesis of DNA is considered a revolution in molecular biology."[47] After six years of research, Temin has shown that a chicken sarcoma virus "generates a DNA replica thanks to one of its enzymes, *inverse transcriptase*; and the replica, according to him, is incorporated into the host cell in the form of provirus."[48] Baltimore drew an analogous conclusion with regard to the leukemia virus in mice. M. Hill, J. Hillova, Todaro, Hatanaka, then Ross, Aviv, Beljanski have published results pointing in the same direction. This research shows that "there exists a molecular mechanism which under certain circumstances, conveys outside information to the organism and inserts it into the DNA of the genetic code."[49] These discoveries cannot yet be interpreted unambiguously. Father Grassé himself writes that "one should not have too many illusions about the value of the interpretations given to them."[50]

Surprisingly, H. G. Cannon cites none of these authors and discusses none of their discoveries in his ardently neo-Lamarckian book.[51] We shall put forward a few commonsense arguments that are perhaps not quite scientific: It is difficult to conceive of DNA as a hereditary program valid "forever." There is something shocking about it, like the homunculi, preformed "forever." It is difficult to go along with the idea that DNA is sheltered from every possible modification originating from the cytoplasm. The transmission of acquired characters is truly a story that is "terminated" and yet "interminable."

For Lamarck, acquisitions are modifications that are caused by factors external to the organism and bring pressure to bear on it through circumstances. They put it in peril and therefore create a need that sets internal powers (fluids and forces) in motion. By their intermediary, a rudimentary organ eventually appears, which frequent use and habit develop. As we have seen, the concept of acquisition became more precise as a function of cellular discoveries in Giard's work at the turn of the century; at that time, "somatic elements" were distinguished from "reproductive elements." The acquired was defined as a somatic modification sufficiently profound to cause an alteration of the reproductive elements. The difference between soma and germ plasm was taken into account even though it had its origin not in Lamarck but Weismann. Tetradactyly by amputation was a somatic property and not a transmissible acquisition; hexadactyly from birth was blastogenic. The thesis was therefore that some changes of the soma become blastogenic. In other words, far from all acquired characters were transmissible. Lamarck had said as much, but it was not superfluous to repeat it in light of Weismann's confused arguments. As genetics grew, the term "acquisition" came to designate all that emerged from modifications external to the gene. When molecular biology and biochemistry discovered DNA, the irreversibility of the message DNA → RNA → amino acids would be discussed. Should this irreversibility be assimilated with the innate? Are all reactions in the opposite direction to be associated with the acquired, be they phenomena observed by Temin, Baltimore, or whomever? These phenomena are extremely complex; they pertain to the internal environment. To refer to acquired characters in this context runs the risk of confusion. What is innate? What is acquired? At this level, the superposition of a controversy about the exact function of DNA onto a debate about innate versus acquired characters seems to me very perilous. Wintrebert has tried to overcome this confusion with a distinction between what one creates and what one undergoes. Are these concepts biological or psychological? If what one creates results in the appearance of a gene, are we not back in a vicious circle, with the gene carrying us back to DNA, to RNA, and so forth, and to the ambiguities mentioned above?

To adopt the terminology of the population geneticist Albert Jacquard, it is tempting to leave the realm of objects[52] for that of factors,[53] which brings into the picture calculations that include the gene as only one of the factors in the model. It is true that Waddington[54] thought it possible that a character acquired by the members of a population might tend to reappear more and more preferentially. But here, he was standing at the level of the gene as object. According to considerations that Albert Jacquard kindly brought to my attention,[55] the gene enters as a factor in various formulas that bring to light the impossibility of raising the issue of the acquired. If one starts from a Mendelian perspective, the issue is not to identify the heredity of an individual I (C I = characters of the individual) as a function of its parental gametes ($\gamma f'$ and $\gamma m'$) and the environment (E), but rather to connect the gametes of the individual with those of the parents. The relation

$$C\,I = F(\gamma f'\ \gamma m'\ E)$$

would then become

$$\gamma\,I = \phi(\gamma f'\ \gamma m').$$

"The function F is interesting, to be sure, but it does not enter directly into the history of the gametes; likewise, the formula for the environment does not appear, which is another way of stating that there is no heredity of acquired characters, contrary to what Lamarck, Darwin, and many others thought." Now it so happens, writes Jacquard, that cytogenetics and molecular genetics have made possible the identification of Mendelian "factors" with the DNA molecules that make up the chromosomes. The issue of the acquired is apparently no longer a question.

The same study draws a similar conclusion for the "additive model." To establish a general cause as a resultant of component causes, it is necessary to add these components together and therefore to assume an additive model. The result can "be analyzed into terms added to each other." One example of the confusion generated by overlooking this issue is precisely the problem of innate versus acquired characters. One can, according to Jacquard, use the term "innate" to refer to genetic information and the term "acquired" to signify the set of the other factors that shaped the individual. The relation between them is expressed by the equation

$$C = F(G, M)$$

where C is the character, G is the genetic information, and M the environment. Is the function F additive? This is virtually impossible to establish, except for variations between each of the two factors from one individual to another; and even then, these variations can only intervene on a very restricted scale. In other words, the problem of the innate and the acquired cannot be raised in general terms.

In conclusion, our rapid survey of these historical examples confirms the announced theoretical difficulty. The transmission of acquisitions, which was fundamental but unquestioned at the time of Lamarck, turned into the inheritance of acquired characters, about which questions were raised until it disappeared almost completely, canceled by the independence of the germ plasm from the soma. Thereafter, the heredity of the innate, which was turned into a dogma by the neo-Darwinians, invalidated Lamarck's first law, while being consistent with the Darwinian principle of natural selection. Lamarck's principle of use and disuse then faded into the background, this time not because it favored evolution, but because it was linked to the "acquired." All the working concepts of the two great laws of the *Philosophie zoologique* were thus displaced and their structures transformed. At the same time, the concept of acquisition changed according to whether it was being discussed at the level or organisms and forms, at the level of the cell, or at the level of the molecules (DNA, RNA, amino acids, etc.).

The field of this problem is presently beset by much uncertainty. Attempts at synthesis like those of Waddington (who takes population genetics as his starting point) avoid the alternative between genetics and the environment by taking refuge in cybernetics. Dobzhansky, Julian Huxley, Gaylord Simpson, J. B. S. Haldane all adopt new perspectives that avoid references to Lamarck versus Darwin.[56] Finally, ideologies have infiltrated the debate. Stalin's policies intervened very violently in the Lysenko affair in 1948. Dominique Lecourt has analyzed this episode,[57] in which it is possible to discern the various factors that brought about Lysenkism: agronomic needs and difficulties, conflicts between the central power and the apparatus of agricultural social layers; the deformation of dialectical materialism that led to the astonishing theory of the two sciences, proletarian and bourgeois, and specifically to the opposition between a so-called bourgeois Mendelo-Morganism and a Michurino-Lysenkist theory of acquired characters mistakenly associated with Lamarck's theory. This Lysenkist tendency allegedly continues to make converts. As far as we are concerned, let us retain the distinction between nonhereditary acquisitions (which sometimes seem to be hereditary because the descendants of the first generation with a given acquisition display the same modifications, but only as a result of being in the same environment!) and the true acquired characters, which are preserved in the descendants even when the latter are returned to the original environment in which the modification had not yet appeared. Lysenko completely shattered the true meaning of the acquired by requiring the condition of a unique environment, namely, the one that brought about the changes. No transmission was ever demonstrated. His extravagant definition of heredity displays the measure of its confusion, if not nonsense: "Heredity is the *property* that a body has of *requiring* specific conditions to live and develop, and of reacting differently to such and such conditions."[58] Lysenko had accused Weismann of racism for his soma theory, and in addition had equated Morgan's thought with Weismann's.

Recently the French daily *Le Monde* has fortunately made the public aware of the "ideologies of scientists," those that are racist, antidemocratic, and elitist unegalitarian and also their opposites. Such ideologies are nothing new. After all, the past offers instances of the defense of "the line" and of "blue blood" on the basis of the innate. Jensen and Herrnstein in the United States, Eysenck in Great Britain, J. P. Hébert in France all represent this position, with its indefensible arguments. *Le Monde* of 30 March 1977 published studies by Albert Jacquard, J. P. Serre, John Stewart, R. Zazzo, and others on the relation of heredity to the environment and the chances of curing genetic diseases. In November 1977 *Le Figaro* saw fit to enter the arena and took Jacquard's studies to task by assimilating his statements to political positions.

Lamarck's self-proclaimed successors have taken us far afield. Have they transformed the problem of the inheritance of acquired characters to any great extent? First, have they changed the content of the concept of acquisition to any great extent? When Albert Jacquard speaks of the acquired in terms of a "lived adventure" of the individual or in terms of "environment," we still seem to be close to Lamarck. But when someone refers to a "Lamarckian scheme" in the work of a given scientist because he notes that cellular differentiation is caused by the mutual contact of cells, or because their identical genetic patrimony does not account for the differences, this is an altogether different story.

In chapter 3 we wondered whether the linkage nature-series-circumstances-anomalies-acquired variations-hereditary transmission of the acquired was a rigorous, scientific, logical chain. We can now answer that Lamarck used concepts with a profound ideological content, but that he placed them on a battlefield where science was still at stake and that he forced these ideas to play the role of concepts. They look like soldiers ill prepared for the battle; but just as the cast leaves the stage at play's end, the nature that reveals the divine plan and the series that translates this plan nevertheless did dislodge creationism before disappearing in turn. The acquired variations and their transmission undergirded transformism before Darwin allowed the evolutionary perspective to dissociate its destiny from the demonstration of acquired characters. The latter then continued to haunt some scientists, but evolution could no longer be doubted. Thus we sometimes see ideas that are later considered unproved help to rid science of archaic notions. At times such as this, one observes the surprising phenomenon of a true idea advanced on dubious grounds. In the apt words of Canguilhem, "Can one not hold [the view] . . . that the progressive production of new scientific knowledge requires, in the future as well as in the past, a certain priority of the intellectual adventure over rationalization . . .?"[59] It is to this "intertwining of ideology and science"[60] that we owe the peculiar flavor of the system of Lamarck, the heir of the Enlightenment and the founder of a new science.

Far from gaining in precision, the concept of acquisition unfortunately became progressively more obscure as its history unfolded. Curiously it seems to have offered no resistance to its passage through all the disciplines. Although emptied of its original meaning, it paradoxically continues to live on in spite of everything.

5 *The Tribulations of the Concept of Adaptation*

There are several reasons for this chapter title. Lamarck does not use the term "adaptation"; the word unquestionably denotes a Darwinian concept. Lamarck's successors nevertheless always considered the first law of the *Philosophie zoologique* to be a law of adaptation. Thus in a lecture of 6 December 1898 Giard formulated "two fundamental laws" and said of the first, "It is the law of ethological reaction or law of adaptation." The famous illustrations of the mole, the aspalax, the proteus, the giraffe, shore birds, and so forth, were adopted by Darwin himself, then by all the neo-Lamarckians and great biologists like Lucien Cuénot as indubitable examples of adaptation. They therefore all tacitly thought that adaptation played a role in Lamarck. But more demanding analysts have examined the matter with greater care. In his book on natural selection, Camille Limoges has formulated the question in exemplary fashion, even though we do not agree with the way in which he answered it: "Although Lamarckism, when confronted with Darwinism, is nearly always presented as a system of thought in which the problem of the transformation of species merges with that of adaptation, there is no problem-set (*problématique*) of adaptation in Lamarck."[1] Moreover, according to the dictionaries, the term appeared in French in its biological meaning only after 1850, as a borrowing from English. The concept of adaptation does not exist for Lamarck. There is, therefore, a blatant contradiction between the abscence of the term and of the problem-set of adaptation, on the one hand, and, on the other, the unanimous attribution of adaptation to Lamarck by Lamarckians who sometimes go so far as to think that the term is even functional in his thought. But the term plays no such role.

Where did the term come from in the France of the late eighteenth century? In its biological meaning, it is not found in the *Dictionnaire de l'Académie française de l'an VIII,* which Lamarck could have consulted. The Littré dictionary of 1874 states that the word is infrequently used. This term, which Darwin imported from a notorious current of natural-theological thought, was not known to Lamarck. On the other hand, Adam Smith's economics includes a theory of adaptation that would become "one of the most important in all of political economy. It would be adopted almost without modification by all the economists who followed Smith . . .".[2] In Adam Smith, this theory concerns the relation of supply to demand, the relation between the need for workers and the rhythm of production, and finally the function of money. The Physiocrats were for the most part contemporaries of Adam Smith, and they were, moreover, fervent believers in the "natural order"—an eminently theological concept, especially in Mercier de la Rivière, who curiously attached himself to Malebranche. Now this "natural order" is the place where theology and economy conjoin to establish in the world a set of coadaptations (although this term is not used.) Smith first became officially known in France in 1777, the date of a reference to him in the *Journal des Savants*. Four different French translations of the *Wealth of Nations* were published between 1779 and 1802. The *Traité d'économie politique* (1803) of J.-B. Say spread his ideas. Adam Smith certainly refuted the Physiocrats with success. But the optimistic idea of a natural order and a harmonious world, religious as well as economic, lived on.

Although the term "adaptation" did not have much currency in France, even in economy or theology, the idea of mechanisms by which the phenomena of nature or of human production finally reached equilibrium for the greatest good of all was in the air; it was believed that, when the order was disturbed, a modification intervened that restored equilibrium and revealed the hand of Providence and the wisdom of the Supreme Author. Thus an idea that designated a particular natural economic phenomenon—for in classical economics the laws of economy are natural and not historical—would give rise to a new set of questions after being transformed by its introduction into biology and reaching the status of a concept. From this point of view, Limoges is right when he says, "When a problem-set exists, the concepts to think it out have already been given, and the concept of adaptation does not exist in Lamarck." And yet we cannot simpy do away with the near unanimity of biologists and historians of science who firmly believe in the existence of adaptation in Lamarck. Not that the argument from authority is decisive—but this unanimity raises a question. If the mistake was made, why was it made? If there is no concept, no theoretical level, no problem-set of adaptation in Lamarck, might there not at least be an area that encompasses the range of a phenomenon, of an object of description before becoming a building block of conceptualization? Once again, we can only appeal to the text. Where in the chain of Lamarckian themes might one find something reminiscent of adaptation?

To recognize this place, we need at least the formulation of an idea, if not the definition of the concept. This formulation borrows from etymology, from the language of the culture, not from a specialized field. In his books *L'Adaptation* and *L'Evolution biologique,* Lucien Cuénot wrote, "The word "adaptation" comes from the Latin *adaptare,* composed of *ad,* 'to' and *aptare* 'to adjust'; it really signifies, as etymology tells us, an adjustment, an assimilation of the organism to the internal and external conditions of life, an adjustment such that the living machine can first of all function, then endure, and finally reproduce itself. This term, which refers exclusively to life, contains the idea of fitness more than utility or necessity."[3] This definition concludes as a function of what follows in the text: "It is absurd to claim that pumice is adapted to floating on water, for pumice stands to gain nothing from a property that bears no relation to the existence of the mineral." Fitness seems to be connected to utility, but not to necessity, insofar as it is not implicated in the laws that govern living species: A given animal may die if it does not adapt. The foregoing definition is too general. Thus the character formulated by Limoges[4] is not necessarily present in it. Limoges states that "in fact there is no adaptation in Lamarck because there is no initiative." We shall not discuss here the merits of this interpretation of Lamarck, but only the validity of the relation between adaptation and initiative. The adjustment, the assimilation mentioned by Cuénot may perfectly well be the result of an action undergone by the living being. Limoges links adjustment to initiative: "If by *adaptation* is meant an adjustment of the organism to its environment that is *effected by the organism itself*, there is no adaptation in Lamarck."[5] In contrast, Yvette Conry writes in reference to Lamarck, "It is of course possible to discuss whether it is a case of adaptation-initiative,"[6] thereby implying that there are other types. Moreover, if one compares Darwinian adaptations with what Giard calls the "law of adaptation" in Lamarck, the semantic contents of the two words are very different. It is therefore necessary to begin with this general meaning as a common denominator.

At what stage of Lamarck's thought does this definition appear? It cannot be at the level of the series, where no modification of the plan of life is at issue. It can only be at the level of circumstances, which force the living being into modifications that will ensure its survival. Giard was therefore correct to call the first law of the *Philosophie zoologique* a law of adaptation, if one adopts the general meaning of the term in ordinary language.[7] Adaptation is then found precisely at the point where organic modification results from the effects of actions and habits. But in a more encompassing and correct sense, it begins at the point where circumstances modify needs, which in turn modify actions and, if the latter are repeated frequently and for a long time, habits as well. Adaptation then occurs precisely at the point where habits modify the organism. By this view, adaptation is a complex act encompassing a cascade of adjustments, whose mechanisms we have not yet attempted to determine. The adjustments bring into play the action of the moral on the physical, the action of the function on the organ. In nature, it takes time for a modification to be inscribed on an organism. Adaptation is a function of a certain duration. "Wherever anything lives," Bergson would later write, "there is, open somewhere, a register in which time is being inscribed."[8] Adaptation is indeed produced in the series, but it only introduces an irregularity, an anomaly, since the essence of the series is its normative regularity. And yet adaptation is not limited to a disturbance of the series; it must also use materials foreseen in the series, and must alter them. It is therefore dependent on the series as well as on circumstances. In more rigorous terms, it is a product of the denaturation of the series by the circumstances.[9]

A closer examination of the phenomenon that seems to be included in the term "adaptation" reveals that the process occurs not directly, but indirectly. At one end of the chain stand circumstances, the "environments" *(milieux)*,[10] the various factors that act on one point of the organism from outside that point; at the other stands organic modification (creation, disappearance, development, degeneration, etc.). Between these extremes lie mediations. This is where the linkage of need, action, and habit comes into play. One should recall the mischief that the substitution of desire, even of will, for need had caused for Lamarck's theory following the publication of Cuvier's eulogy. In the *Philosophie zoologique,* the concept of effort intervenes between needs and organic modification.[11] Lamarck borrows many examples from human activity (transported grains, cultivated plants, domesticated animals) or the paths of nature (the whale, the mole, the aspalax, reptiles) to illustrate the disappearance of unused organs. He mentions tree and shore birds, the giraffe, and so forth as instances of the development of organs by use. The central concept of use is precisely the type of action stimulated by needs, which are in turn stimulated by circumstances. But what then is the mechanism that comes into play between habitual action, constant use, and the organ? "In the second part [of the *Philosophie zoologique*]," writes Lamarck, "I will show that when the will determines an animal to a given action, the organs that must execute this action are immediately stimulated by the affluence of subtle fluids (the nervous fluid) that become the determining cause of the movements that the action in question requires."[12]

The mechanism of the fluids is summarily explained in chapter XI, but sufficiently so to rule out a vitalistic interpretation. The "exciting cause" of fluid motion, the "spring," the "force" that sets off the motion is a physical force.[13] Lamarck places much stress on the danger of vital principles. "Whenever we abandon nature to give way to the fantastic leaps of our imagination, we lose ourselves in vagaries, and our efforts result in nothing but errors."[14] Even if we could not discover the cause of organic movements, "it would nevertheless be completely evident that this cause exists and that it is physical."[15] Moreover, the examples of caloric and the electric and magnetic fluids are physical in character; the nervous fluid is of the same nature. The fluids are set into motion by any pressure on the more solid parts—the former, being unstable, are agitated by the slightest cause. Caloric and electricity are two essential causes of motions and the resulting changes: "In animals whose organization includes few components, the caloric of the environment seems to be of itself sufficient for orgasm and irritability in these bodies. . . ." There is indeed a vital property, irritability, but it does not proceed from a mysterious principle. Instead, it is a specific effect of internal caloric. At the time of Lamarck, the preeminence of caloric was such as to cause hesitation about the nature of heat: Was it a fluid or a molecular motion? Carnot was still in doubt; Lavoisier had mentioned both options. Only toward the middle of the following century would thermodynamics become an established field. Here again, Lamarck was no precursor. He discussed the cause of life in the ancient terms of the subtle fluid; but when he discerned the importance of the "caloric" and "electric" and "magnetic" fluids, he was on the road that later would lead to the experiments on the origin of life.

The phenomena that explain the influence of the environment on the organism are therefore physical in nature, and yet the living being alone reacts with modifications. The fluids are in us and outside us; they create in the living being, and only in the living being, "a particular tension[16] so active" that it makes the flexible parts of an animal "susceptible to sudden and instantaneous reaction."[17] This sentence identifies the part of the reaction that would later be termed "adaptation." This process makes explicit what Lamarck meant by "life,"[18] the new concept that is the object of biology, and which not only he identified. But what was life for Lamarck? The answer is of some importance for the concept of adaptation. In 1820, at the end of his career, Lamarck wrote, "Life, in a body whose order and state of affairs can make it manifest, is assuredly, as I have said, a real power that gives rise to numerous phenomena. This power has, however, neither goal nor intention. It can do only what it does; it is only a set of acting causes, not a particular being. I was the first to establish this truth at a time when *life* was still thought to be a *principle,* an *archeia,* a *being* of some sort."[19] So he concluded a line of research that he had begun clarifying around 1800. Note the following text from 1802: "I am convinced that life is a very natural phenomenon, a physical fact, in truth a little complicated; it is not any *particular* being."[20] His new definition reads, "Life is an order and a state of affairs in the parts of every body that possesses it, that allows or makes possible in it the execution of organic movement and that, as long as they exist, effectively oppose death."

Is there a change in Lamarck's conception between the *Considérations sur l'organisation des corps vivants* and the *Histoire naturelle des animaux sans vertèbres*? Burkhardt[21] thinks that Lamarck appealed to a mysterious vital principle in his earlier writings, and he dates this position to 1794. If this is true, between this date and 1802 he came a long way. For in 1802, as noted, he was already writing that "life is a very natural phenomenon, a physical fact,"[22] and in his *Philosophie zoologique* of 1809 he confirmed that "all if this is only a matter of purely physical phenomena."[23] We already quoted the remarkably contemporary text from the *Système analytique* on the absence of teleology in life. Many other texts could be cited to show that if there ever was a vitalist temptation, it was very quickly replaced by a logic of life of a physicochemical sort, albeit a very specific one. If we maintain with force that Lamarck holds fast to two positions that he considers obvious, namely, (1) the specificity of life, which goes hand in hand with the term "biology" applied to the new science, and (2) the integration of life into the physicochemical realm, then faced with the impossibility of joining these two data, we will almost accept Wintrebert's thesis: "One examines the way Lamarck proceeded; one notes that he could not do otherwise. He lacked the necessary biochemical knowledge required to rest his theory on solid foundations. He could convince his contemporaries only by [appealing to] the correctness of the convergence of a necessity imposed by the environment and of the effective realization of the organ adapted to this environment, without being able to define the link between them by anything but the word adaptation."[24] Here Wintrebert overlooks the absence of the term in Lamarck, but what he says before this blunder helps us understand the absence of the concept and the term.

It is once again obvious that Lamarck was not a precursor. Without being able to explore it in depth, he nevertheless pointed to the place where the series of operations takes place that would establish between the organism and the internal and external environments an adjustment that probably includes mechanical regulation, but no teleology. May we call this phenomenon "adaptation?"

If Camille Limoges rejects this view, it is because for him adaptation refers to the Darwinian concept. Let us leave aside for the moment the relation between Lamarck and Darwin and retain Lucien Cuénot's definition. Limoges's objections are reducible to five points:

1. Neither the term nor the concept are in Lamarck; without a concept, there is no problem-set.
2. The organism undergoes adjustment; it displays no initiative, and hence there is no adaptation.
3. The locus of adaptation cannot be the series, which follows a progression based on its own internal necessity, and would function even "in spite of the environment."
4. The locus of adaptation cannot be the circumstance that engenders the anomaly, and therefore operates at a pathological and negative level; it does not condition the progression of living forms.
5. The species are adapted in advance, since they cannot disappear; for Lamarck there are indeed no extinct species. It is well known that he does imagine the extermination of certain animals by man, or unknown regions where fossils live on, or else a transformation of fossils into actual species. If the impossible extinction of some species is due to a "natural harmony," then life becomes an organizing power, and is "not simply a hydraulic power," as Jean Rostand rather amusingly puts it.[25]

We have already given our opinion on the first point: Before a concept appears in a language or a scientific theory, it may obscurely haunt the neighborhood of the phenomenon it will later be able to explain. In this case, the concept is certainly not a ghost about to become incarnate, but rather the act of thought in genesis.

The answer to the second issue is more complex. Wintrebert has put forward an idea of creative adaptation, at the level of an unconscious power that would lead to the emergence of new genes. When Yvette Conry wonders whether Lamarck presents an instance of "adaptation-initiative," she thereby implies the possibility of other modes of adaptation. In both cases, for opposite reasons, Lamarck is alleged to use the phenomenon of adaptation, but no account is given of the absence of the term. There is a way Lamarck used "initiative" that might suggest "adaptation": The use or nonuse of an organ would depend exclusively on the "inner feeling" (*sentiment intérieur*), the will, consciousness. This is what Darwin thought, and all those who followed him[26] in reading Lamarck from a factor he had invoked, but that could play a role only by means of need, action, and effort, all of which bring fluids into play. I do not think that the problem is fundamentally one of hydraulics. It lies rather in the following question: What is the specific trait called "life," such that fluids have an effect that they display under no other circumstances, namely, triggering an answer that goes beyond the terms in which the question is phrased? There must be something that reacts to the exciting cause, by producing an effect that this cause did not carry within itself. Cuénot and the philosopher Edouard le Roy thought of this as the power of "invention"; Wintrebert considers it the "creative" power of life. This effect cannot be compared to the displacement of a ball pushed by an external mover. Transformation is not displacement. These two types of motion cannot be reduced to each other. To conclude this second point, it should be pointed out that the concept of initiative is difficult to circumscribe. On the one hand, it cannot be completely excluded once one takes the specificity of life as a given; on the other, it is not certain that adaptation and initiative are linked. In other words, solid though it be, C. Limoges's analysis leads to excessively categorical conclusions.

Third, we have agreed that adaptation is impossible at the level of the series. The latter is driven by an internal teleology that escapes us, since it comes from the "Supreme Author" of nature. We can read it only when it is translated into the language of efficient causality. Since the series is "what ought to have been," by definition it does not adapt. And yet if the series does not end, is not destroyed, does not break when faced with a circumstance that hinders its development, it is because the series has the wherewithal to undergo alterations that certainly denature it, but also allow it to continue its march forward at the price of this deformation. Here too the issue is complex.

Granted that the circumstances produce alterations, anomalies, and so forth, is this a reason to concede the fourth point and reject every sort of adaptation phenomenon? We have already said what we thought of the double character of the effects of changes in, for example, environment and climate. The pathological character of the effects of circumstances has also been mentioned; in more precise terms, one should speak here of teratology. But this monstrosity with respect to design essentially becomes the diversity of beings before our very eyes and ensures life's triumph over death. What then does the rejection of the term "adaptation" imply, if not once again a reference to Darwinism? Limoges's point of view is legitimate, if one concedes that the only viable meaning of adaptation is Darwin's. But if one starts from a more general meaning, it is indeed possible to speak of adaptive phenomena in Lamarck. It might be objected that mechanism is not analyzed scientifically and hence that to speak of adaptation is to introduce an unacceptable term. This would require an examination of the Darwinian mechanism itself. Was it possible for Darwin to give a completely scientific content to adaptation? Where is the issue today? We turn to these questions in the remainder of this and the following chapter.

Finally, there remains the long story of the extinct species, which for Limoges involves "the opposition between the relation of the living to the inert in Lamarck and the Darwinian concept of adaptation."[27] In Limoges's interpretation of Lamarck, every struggle, every extinction depends on the "natural economy"; his hostility to the extinction of species fully displays his agreement with the theory of "the equilibrium of the universe" over which the Supreme Author watches. In this sense, everything in the universe involves a reciprocal adaptation, not in the sense that each being adapts, but in the sense that it is preadapted. There is, in other words, a "natural police," as Linnaeus had suggested.[28] The concept of natural economy is precisely what Darwin would eliminate. A defective adaptation would eliminate the disfavored species; they die like individuals that old age has gradually made unsuited to adaptation. In this argument, Limoges unambiguously clarifies Darwinian adaptation in relation to Lamarck, for whom there seems to be nothing but harmonies that the scientist must read though a network of causalities, not through the effusions of Bernardin de Saint-Pierre. With regard to adaptation, Lamarck certainly lags behind Darwin. The point deserves to emphasized: We want to show precisely that in 1809 Lamarck was not thinking about 1859. He eliminated creationism, but retained the idea of design and natural economy, the idea of a God of the Enlightenment, the God of Rousseau, Goethe, and—taking differences into account—so many others. What is more, Lamarck used nature, the chain of being, and harmonies against creationism. In other words, he attacked one ideology with another. The first, more archaic, rested on the literalism of the Biblical account; the second, more rational, was based on a God of order, of the unity of design, of a creation expressed by evolution. Yet we have seen Lamarck hestitate between nature as design and life as organizing power on the one hand, and the natural as opposed to the supernatural, and leading to the universe of laws, to physical mechanism, on the other. Darwin would eliminate what remained of religious and even rational ideology. In this contrast, we see the striking difference between Darwinian adaptation and a Lamarckian phenomenon retrospectively termed "adaptive." The latter is still very ideological; the former is on the road to rigorous conceptualization. Adaptation is thus highly suited to covering or uncovering ideological effects by the content one grants the term and by the way one brings it into play between Larmarck and Darwin, from Darwin to Lamarck, and from the "neos" of every stripe to their founding fathers. Adaptation is the driving force of every evolutionary theory; for adaptation is variation, and it is variation that allows living beings to change, by transmitting their modifications to their descendants.

At this point we have reached an impasse: We are developing an argument on the basis of a general, and therefore insufficient, definition of adaptation. Camille Limoges, on the other hand, uses the Darwinian criterion. Since these two positions have different premises, the confrontation is not conclusive. One of the two positions should have uncovered an internal inconsistency in the other. I do not think that this is the case. Both positions hold firmly, if one concedes their starting points. The debate can move forward only by examining the Darwinian criterion put forth by Limoges. By so doing we begin the process that we shall develop more extensively in the following chapter. For now, we shall project the Darwinian criterion onto Lamarck; that is, we shall take Limoges's position. "Heaven forfend me from Lamarck nonsense of a 'tendency to progression,' '*adaptations* from the slow willing of animals' etc."[29] What did Darwin know about Lamarck at the time? He had read the first edition of Lyell's *Principles of Geology*, but had taken no notes on the latter's lenghty critique of Lamarck, which fills one third of the book. He would later tell Lyell (11 October 1859) that he had borrowed nothing from Lamarck. The *Autobiography,* on the other hand, mentions information about Lamarck given to Darwin by Robert Edmund Grant, a friend from the University of Edinburgh. Now in 1836, the *Edinburgh*

New Philosophical Journal[30] had published an unsigned translation of Cuvier's eulogy, which spread the confusion between need and will, and in turn heaped ridicule on Lamarckism. Darwin read and annotated the *Philosophie zoologique;* he thought and wrote that Lamarck had plagiarized his grandfather Erasmus Darwin. But this is only an anecdote. The important point is that the notion of direct dominance of the will over the modification of organs—a view indebted to Cuvier's distortion—always colored Darwin's reading of Lamarck. This approximates the gist of Jean Rostand's remark, alluded to earlier.[31] C. Limoges writes, "Through the Anglo-Saxon intellectual tradition in which he participated, Darwin read Lamarck as an author who discussed the problem of adaptation by presenting an unsatisfactory solution, that of modifications produced by the "interior sentiment" (*sentiment intérieur*). He reads Lamarck like an English naturalist of the first half of the nineteenth century, like a naturalist who has read Erasmus Darwin's *Zoonomia.* And since on this issue we participate in the mental universe created by the Darwinian revolution, for the last century we have been reading Lamarck as Darwin did."[32]

Now Darwin thought he had discovered "the very simple way in which species adapt themselves perfectly to various ends."[33] The concept of adaptation rested on the concept of difference, which Darwin considered the basis of varieties, however slight the difference might be. "The passage from one degree of difference to another can in many cases simply be the result of the nature of the organism and of the different physical conditions to which it has long been exposed. But when more important adaptation characters are involved, the passage from one degree of difference to another can surely be attributed to the cumulative effect of natural selection and to the effect of increasing use or disuse of the parts."[34] Note here the role given to factors that are clearly those of Lamarck (physical conditions, use, disuse), but especially the essential function of the nature of the organism. Darwin could observe the passages from one difference to another in the Galapagos Islands and in his southward travels through South America. It is through them that he discerned the formation of species. Later Darwin himself would connect these observations from his voyage on the *Beagle* to the nonfixity of species: "At last, gleams of light have come and I am almost convinced (quite contrary to the opinion I started with) that species are not (it is almost like confessing a murder) immutable."[35] From the outset, the problem of adaptation was therefore tied to the transformation of species; so it is that varieties emerge from variations and, when the varieties become more accentuated, lead to the appearance of new species. The importance of "adaptation"—the term is Darwin's—in its fundamental relation to evolution may be found in Lamarck with regard to the phenomenon that, it was our impression, would later be called adaptation.

What were the factors of modifications? "The subject haunted me." But it was obvious that "neither the action of the surrounding conditions, nor the will of organisms . . . could account for the innumerable cases in which organisms of every kind are beautifully adapted to their habits of life—for instance a woodpecker or a tree-frog to climb trees. . . . I had always been much struck by such adaptations and until these could be explained, it seemed to me almost useless to endeavour to prove adaptations by indirect evidence that species have been modified."[36] This text is crucial for the purpose of locating the concept of adaptation within evolutionary theory. Darwin expressed the same view elsewhere: "Nevertheless, such a conclusion, even if well-founded,[37] would be unsatisfactory, until it could be shown how the innumerable species inhabiting this world have been modified, so as to acquire that perfection of structure and coadaptation which most justly excites our admiration."[38]

These texts, which truly express the core of the problem, can be fully and theoretically understood only if they are seen from the twofold aspect of their biological scope and their theological source. Indeed, at stake in Darwin's research and his later struggle is the elimination of the concepts of adaptation and coadaptation, which are the foundation of English natural theology, and which he knows all the better for having started out in a career in the church. The equilibrium of life, "the balance of animals," as William Derham puts it in his *Physico-theology,* goes back to the beginning of the eighteenth century. This vision of the universe is very close to the "natural police" of Linnaeus's *Systema naturae.* The latter celebrates "the splendor of the republic of nature," its equilibrium, its proportion. He cites Derham and marvels at the way animals feed upon one another. The division of labor is as remarkable here as it is in the factories celebrated by Adam Smith at the end of the eighteenth century. Capitalist economics and the admirable government of Providence come together and resemble each other. From Ray[39] to Bentley and especially William Paley,[40] all the theologians rest their proofs for the existence of God on natural interrelations, on adaptations—the very word used. The naturalists are very aware of natural theology.[41] Here adaptation is one of the Supreme Engineer's devices and not a biological function. It is not an act, but an aspect of Creation. It is perfect from the origin. It is impossible to speak of a transmutation of species where adaptation is preestablished and overseen by the Creator. Thus theological adaptation blocks the road to evolution. For Darwin, the elimination of this obstacle did not imply the annihilation of the concept, but the substitution of scientific content for its theological character. In the words of Limoges,

In a fixist world, adaptation is only an adjustment to the environment, not an initiative of the living being, nor a molding action of the environment. . . .[42] In a transformist conception, either the environment molds and modifies the organism without any adjusting action on the part of the living being itself (as is the case in Lamarck) or the adjustment comes from the living organisms, and only in this case is it truly adaptation. The latter may come from an effort of the living organism, as is the case with Erasmus Darwin and would be the case with Lamarck, if the "interior sentiment" actually had the importance and the autonomy generally imputed to it. But it can also come from variations of the living being, which are independent of any initiative on its part, and have adaptive value only in an aleatory fashion. This is Darwin's solution.[43]

The text is clear; it confirms Limoges's thesis about Lamarck, a thesis about which we have expressed some reservations. On the other hand, we retain the distinction between the theological and the biological, and the exact formulation of the Darwinian position. The last sentence invites us to determine, through the Darwinian concept of variation, how and to what extent Darwin effects the reconstruction of the concept of adaptation.

The problem of variations, of their origin and their conservation, is by Darwin's own account major: "With respect to the causes of variability, we are in all cases very ignorant."[44] But the very fact of variation tied to difference can be circumscribed; the former includes the latter, but the reciprocal is not true, for variation is a difference that "appeared"; that is, it did not exist at a specific moment and emerged at the next. It implies a change in time, whereas a difference can be original and fixed. "Any variation that is not inherited is unimportant for us," writes Darwin. Behind variation lies variability, that is, the plasticity of the organism, which is accentuated "under domestication."[45] The breeder works by selection from the most variable types, but in nature "we have abundant evidence of the constant occurrence under nature of slight individual differences of the most diversified kinds."[46] In the *Origin of Species,* Darwin attributes these variations to chance. The organism varies, and that is all there is to say; we know nothing whatsoever about the phenomenon. Yet Darwin indicates two sources of variation: the nature of the organism and the nature of physical conditions. As is known, he accepts indifferently the innate and the acquired. He simply concedes more to the innate in the case of slight variations that accumulate with time. These emerge suddenly by chance, that is, without our knowing why, without our being able to grasp the relation between the cause and the effect. They are then called "spontaneous" or even "accidental," but "only in the sense in which we say that a fragment of rock dropped from a height owes its shape to accident."[47] Here he opposes every intentional connotation; he does not immediately see the relation between these "individual variabilities" on the one hand, and on the other, the formation of species and "all those exquisite adaptations of one part of the organization to another part, and to the conditions of life, and of one distinct organic being with another being."[48] He often expresses his bewilderment before the attempt to establish the relation.

It is through the concept of utility that the relation between variation and adaptation is first established. The useful and use must not be confused, however: The first term is Darwinian and designates what is favorable to life; the second is Lamarckian, although Darwin applies it with increasing frequency as he progresses in his research, and designates the exercise of a function and through it of an organ. Every advantage, however slight it might be, brings with it a chance for survival. It is precisely at this point that, through the intermediary of the useful, the variation is connected with adaptation, which is itself tied to natural selection. "This preservation of favourable variations and the rejection of injurious variations, I call Natural Selection."[49] The giraffe appears here as a privileged living being progressively adapted, for Lamarck and Darwin, along with the "tumbler pigeon" so dear to Darwin.

We have now reached the point where it is possible to bring out the entirety of the Darwinian structure and to see more clearly its distance from Lamarck's development. The serial model of the chain dominates Lamarck's thought: circumstances, needs, actions, habits through use and modification of the organs—such is the succession of concepts in his work. For Darwin, thought is an edifice of structures rather than a sequence. Variation is its foundation; its fate is tied to its utility with respect to the environment. Utility ensures the variability of being within the individual and, by transmission of this variation to the descendants, makes it possible for the living being to compete with the all-too-numerous individuals from which the most fit are naturally selected. The opposition of circumstances to variation is manifest. For Lamarck, primacy goes to the variation of the organism. The latter may be a function of either the spontaneous variability of individuals, no two of which are identical, or the variation triggered by the environment. But even as it is being triggered, the variation that testifies to the specificity of life is a foundation.

The second point is the difference between utility and use. Utility leads to use, that is, to the exercise of a function; but the latter does not modify the organ, since modification precedes use. Use plays a secondary role in Darwin, whereas utility is a distinct advantage in the selection process; it is the individual's trump card in its struggle for existence. The last point is the emergence of the concept of "environment" (*milieu*). The environment is first of all physical, but also, and especially, biological. Living beings are intimately dependent upon each other.[50] Darwin does not tire of giving examples of this "ecological" equilibrium (if we may use the term anachronistically) such as the change in vegetation due to the introduction of pine trees, but also the effect of plants on the appearance of insects, and even beyond that of the insects on insect-eating birds. The environment (*le milieu*)—a term not used by Lamarck—is not at all a set of circumstances, but a system that can spill over into another system, without ceasing to function by means of eliminaton and selection, death for some and survival for others. This is not a narrative or descriptive, but an explanatory mode.

It is hardly necessary to add that natural selection, which ensures the transmission of modifications based on the utility of variations that appeared randomly, is scarcely comparable to the struggle that, for Lamarck or his predecessors, leads to a simple decrease in the number of beings that would otherwise increase excessively. In Darwin, selection does eliminate the risk of a large number—the influence of Malthus on his thought is well known—but its function is to ensure the survival of the fittest. He has in mind not cock fights, but the resistance to hostile factors and especially the competition between useful factors and the improvement of living organisms. Selection is, as we have seen, the motor of evolution, and not just a curb on excessively large numbers. It makes the system function through the mediation of life, death, the progression of favorable crossbreeding by heredity. The interplay of a self-regulated system has been substituted for a heroic serial design perturbed by circumstances. Behind the design stood the Supreme Author; behind the system, chance—in other words, as Darwin concedes, our ignorance, but also the dismissal of teleology. Variations are due to chance; they are accidental. Just as the shape of the rock at the bottom of the precipice depends on antecedent causes without embodying the design of some God for the builder who will use the rock, so the variations by which the breeder could obtain domestic species and varieties do not point to foresight. Did the Creator ordain that "the crop and tail feathers of a pigeon should vary to allow the breeder's fantasy to create his grotesque 'pouters' and races of peacocks?" Certainly not; but there remains a difficulty in reconciling the omniscience and omnipotence of God with chance. Thus in these two opposite perspectives, adaptation appears in Darwin as the result of an adjustment mechanism between chance variations and the environment, whereas the Lamarckian phenomenon later identified as adaptation results from a mysterious relation between the design and its opposed perturbation and allows it to be realized only through disfigured and unrecognizable manifestations. Darwin has eliminated the theological and providential content of the concept. Lamarck has described a phenomenon that allows an altered divine plan to survive in the background; he has destroyed fixism and creationism, but has not completely eliminated deism in science.

The opposition between Lamarck and Darwin did not prevent the latter from judging variation to be both essential and insufficient, or from making ever more room in later editions of the *Origin* for the conditions of existence as explanations of "the innumerable configurations perfectly adapted to the vital habits of each species"[51] or the admirable adaptations, of which the giraffe is the most beautiful example. Of course, Darwin realized that in order to be useful and viable, an adaptive variation like the forelegs or the neck of the giraffe presupposes simultaneous variations in other organs. But he thought that if the variations were imperceptible, if they were imagined in a time interval and a number of descendants so great that they defied the imagination, then the objection would be overcome. Characters become additive, and their accumulation allows the game of chance to orient itself automatically in the direction of increasing adaptation. Thus biological adaptation by natural selection has driven away providential preadaptation. The individual adapts; he is not preadapted. Regulation takes place by nefarious elimination (death) and the preservation and transmission of the useful (life). Nature is an enormous breeding ground, where the breeder, if he exists, has no scientific involvement! Heredity still remains obscure; in Darwin's theory, it stands as a blank filled by the conjecture of pangenesis. It was necessary to give an account of adaptation in order to make evolution acceptable, and this required that heredity be better known. When genetics came into being, Darwin was already dead, so that the concept of adaptation was theoretically criticizable, as Yvette Conry has shown in her remarkable book.[52]

Nevertheless a long road was traveled from Lamarck's "story" to the Darwinian "proof," in spite of all the unknowns in it. The gradual elimination of ideology is evident. With Darwin, the entire natural economy, along with its Author, is moved beyond the boundary of science. The notorious problem of extinct species disappears, along with its solution by cosmic cataclysms. This is not to say that Lamarckian factors are not present in Darwin, and the neo-Lamarckians never fail to point this out. But in the latter's work, the "conditions of existence" see their role subordinated, or at the very least tied, to natural selection. If this process were brought into question, then the problem-set would change. And in fact questions are being raised today. The primacy of function over organ regained some credit after the second half of the nineteenth century when Milne-Edwards applied cellular theory to the study of monocells, which he compared to "workers without tools." It is indeed known that in multicellulars the functions that become differentiated at the same time as their organs are due to the activity of the protoplasm.[53] And somewhat later, toward the end of the nineteenth century, the evidence of the "real impact" in botany[54] of external circumstances and conditions of existence is underscored. Here one ought to recall the work in our own time involving the restrictive role of the gene and the role of the environment.[55] One would then be led to conclude from the questions presently raised about Darwin and the resurgence of Lamarckian factors that if Lamarck is not made out to be a more or less presumptuous precursor, he regains his real merits. The reproaches leveled at him after studying him through his successors, and Darwin in particular, are not justified if his explanations are restricted to the "range of possibilities" open to him. One would then concede that "the effort made by the need" signifies for Lamarck "the spontaneity of the living being," its "power of reaction" and its "biological specificity." Yvette Conry, who made this remark, adds, "But if the organism is the center of response and the subject of behavior, has not an 'adaptation' thereby been defined? In any event, the idea is physiologically circumscribed and postulated in Lamarckism, whereas in Darwinism, the concept is zoologically determined and questioned, so that it gives rise to a problem-set."[56] In this case, with due respect to Camille Limoges, several senses of the term "adaptation" must be conceded, and even the thing itself when the concept is not there. François Jacob does not hesitate to refer to Lamarckian adaptations three times on the same page. Let us quote this passage in particular: "This constant interference between the faculties of the organism itself and the external circumstances results from what Lamarck considers to be one of the most indisputable properties of living beings: their power of adaptation to the conditions of their life, the concordance between the organism and its surroundings."[57] By emphasizing that this concordance is a harmony of nature, F. Jacob locates adaptation at the ideological level where living beings "are adapted." But when he says a few lines later that "the environment adapts, in contrast to heredity, which creates," he locates adaptation at the level of the circumstances, and not at the level of the divine plan. The fact is that the Lamarckian sense of the term stands between the two, thereby justifying Y. Conry's distinction between what is called adaptation in discussing Lamarck and what Darwin calls adaptation: "In short, Darwinism is an etiology of evolution . . . that begins with the problem-set of adaptation, with the conceptual norm of selection (which brings an ecological context into play) and the assumption of variation."[58]

We conclude that in one sense, Darwin did not succeed in making adaptation anything more than a brute fact, that he did not succeed in theorizing, even though he went much further than Lamarck—as is indeed normal if Lamarck is not forced to be a precursor. A broader study of what scientists and thinkers in the second half of the nineteenth century made of Lamarck through Darwin—a function of what they made of Darwin through Lamarck—will confirm the unusual tribulations of the concept of adaptation. Although it is not completely justified from a scientific point of view and is exposed more than any other notion to ideological contamination, this concept is the privileged vantage point from which the relation between Lamarck and Darwin can be seen, with all of its clarities and obscurities. The final word belongs to the author of that remarkable book on Darwinism in France, Yvette Conry: "In all honesty, is the concept of adaptation so clear to contemporary minds? Does it not still today serve as a refuge for more or less explicit systems of metaphysics?"[59]

6 *When History Is Read Backward*

At this stage, the mixture of ideological and scientific elements no longer needs to be demonstrated. If Lamarck the scientist from the beginning of the nineteenth century is today unrecognizable, it is certainly due in part to all the intervening scientific discoveries. The questions asked today are completely different. But it is also due to the fact that religious, political, and ethical ideologies have splashed about to their hearts' content in the history of science, and are inseparable from it. Lamarck the scientist was not only imbued with one or more ideologies; he has also been saddled with ours and the intervening ones.

Thus in order to see our own time in all of its historical depth, and not only with the superficiality of a certain brand of structuralism, we shall proceed upstream through the history of science. To clarify: In the history of science, as indeed in history generally, there are "hot periods," for example, successive waves of neo-Darwinism or neo-Lamarckism. In the latter case, the turn of the century and the present are such periods. One day, someone decides to go back to the sources, like Yvette Conry in the book to which we have frequently referred, where the relations of Darwin and Lamarck color Darwin's reception in France. We propose to retrace our steps toward Lamarck beginning from the period extending roughly from 1860 to 1910. We shall then make a second start in the contemporary period in order to see how the various strands of neo-Lamarckism are integrated into a retrogressive analysis leading from the neo-Lamarckians to the Lamarckians, and then to Lamarck himself. The period between the Lamarckians and the neo-Lamarckians saw the appearance of Darwinian theory and several discoveries, the most important of which are cellular structure and genetics. Some scientists claimed to be Darwinians, but in fact introduced Lamarckian doctrine into their procedure. Others, who as scientists were rather more favorable to Darwin, identifed themselves as Lamarckians for ideological reasons. The situation is complex. Thanks to this analysis, we hope to find an answer to the question, How did the myth of Lamarck the precursor come into being? At this point that sort of speculative fascination seems to come to the fore by which the scientist, philosopher, or theologian—sometimes in three persons, sometimes in one—thinks he sees in Lamarck the prelude to his own thought. To proceed retrogressively is properly to trace the production of this illusion to its source. Everyone, in a given period, has looked at himself in the mirror of an unreal Lamarck, whose broad views were suited to the task of nourishing a variety of later productions. This mirage will perhaps disappear if for the phantasmagorical product of the past by the present we substitute a relation of the past to the past, that is, of Lamarck's present to the present of his environment. This is what Burkhardt has done in *The Spirit of System,*[1] and what Limoges, in his review, calls "Lamarck and his environment." But before beginning the analysis, let us first identify the period we want to reach.

We shall not pause over those of Lamarck's contemporaries who more or less followed his doctrine: Poiret, who collaborated with him on the *Dictionnaire de botanique de l'Encyclopédie méthodique,* or Etienne Geoffroy Saint-Hilaire, who thought of himself not as a disciple but as a colleague, even when he defended transformism and its founder.

During Lamarck's productive years and in the few years following his death, three of the greatest thinkers of the period followed with consuming interest the emergence of biology: Goethe, Schopenhauer, and Auguste Comte. As a partisan of the synthetic method, broad views, and grand syntheses, Goethe supported with enthusiasm the thesis of Geoffroy Saint-Hilaire against Cuvier before the latter's death in an account of the 1830 debate between the two scientists before the Académie royale des sciences. He mentioned the great event to Eckermann, who thought Goethe was referring to the July revolution. Said Goethe, "We do not appear to understand each other. . . . I am not speaking of those people, but of something quite different. I am speaking of the contest, so important for science, between Cuvier and Geoffrey de Saint-Hilaire [*sic*], which has come to an open rupture at the academy."[2] He was thinking of course that Geoffroy Saint-Hilaire's work supported his own work on the intermaxillary bone. This still does not explain his complete silence on Lamarck and transformism, especially when one sees his attraction to evolution and considers the fact that Lamarck had just died.

Schopenhauer wrote *On the Will in Nature* while he was retired at Frankfurt-am-Main and cut off from the mainstream of German philosophy. The book, which was published in 1836, aimed to connect his metaphysics with the science of the day. For him the philosopher and the scientist were like two miners unknowingly walking toward each other in a mine shaft. In the chapter on "comparative anatomy," he paid a vibrant tribute to Lamarck, but thought that the transformation of species in time was a "mistake of genius."[3] In his view, modifications caused by use seemed to him to depend on the will. Had he too read Lamarck through Cuvier? For Schopenhauer, will stood outside of time. His thought confronted that of Lamarck, but was not influenced by it.

Completely different in character is the relation of Auguste Comte to Lamarck, by way of H. Ducrotay de Blainville, a professor at the Museum. In the lectures of his *Cours de philosophie positive* and in the *Système de politique positive,*[4] Comte assigned such a prominent role to biology that Canguilhem thinks it had an even greater importance to Comte than sociology. The two great concepts Comte borrowed from Lamarck were those of "environment" (*milieu*) and series. His rejection of teleology confirms, if need be, his positivism. One should note the singular environment, which here replaces environments (*milieux*) and circumstances well before Darwin; one should note also the call for a real theory of environments. But at the same time Auguste Comte wanted to prevent Lamarck's theory from serving as a support for mechanistic monism and the eventual reduction of life to inorganic elements. In Comte's view, life had its own initiative, its own specificity; it could not be integrated into a cosmological imperialism. This position bears the imprint of Barthes and of vitalism. Comte welcomed the transmission of acquired characters. He was pleased that it accorded with the notion of progress, but he would not turn it into the basis of a genealogy of living beings. He conceived of progress in terms of the mastery of matter by the living. Liberated from teleological theologism, man would impose his own ends on nature. By defending the living, Comte ensured the sovereignty of the highest form of life, mankind; and the series, seen as a degradation starting from mankind, facilitated the transfer of biological concepts to sociology. Here one might concede a use of Lamarckism and the rudiments of a scientific ideology. In my opinion, however, Lamarck's thought was absorbed into Comte's powerful synthesis rather than the other way around.

All of this suggests that between 1829 and 1850 Lamarck was not as unknown as he is so often made out to be. Philosophical minds were sensitive to the breadth of his "philosophy."

We shall spend no time on such Lamarckians as Bory Saint-Vincent, the coauthor of the *Dictionnaire classique d'histoire naturelle,* or Lecoq, the author of a good study of geographical botany. We shall leave Naudin to his vacillations between Darwinism, teleology, and "creative power."[5] Let us turn immediately to the neo-Lamarckians, whose growth was due to the fact that Darwin had brought Lamarck back into the limelight. It was around 1870 that the neo-Lamarckian fireworks went off, with effects that were to last nearly forty years.

It is important to demarcate the field that was delimited and structured by Darwinism, for it was in this field that there took place the event whose extreme character probably derived from scientific patriotism related to the defeat of 1870. Did Darwin control all of the strongholds of biology? England had welcomed his theory; Lyell, Wallace, Hooker, Huxley, and Spencer all gravitated around him. Italy and Switzerland were sensitive to his influence. But it was especially in Germany that he carried great weight. France, by contrast, resisted or deformed the Darwinian contribution: The first reference to Darwin dates to 1860. The first French edition of the *Origin of Species* came out in 1862. Only in 1878 was Darwin elected a corresponding member of the Académie des sciences. The *Dictionnaire de l'Académie française* first included an article "Darwinisme" in 1875. Flourens had been a Darwin opponent; Janet, Renouvier, and Cournot took position against him as well.[6] In spite of this reticence and these resistances, Darwin gradually became known. Although he was the leader of the Lamarckians, Giard campaigned for him. Ideologies seized his theory. Vacher de Lapouge unleashed a scandal at Montpellier by adducing Darwinian factors in support of his racism. "Foreign conclusions are attributed to the doctrine."[7] Even before they had been analyzed scientifically, the concepts of "natural selection" and "struggle for existence" were exploited outside the biological realm. Conservatives and clericals sensed a doctrine dangerous for a natural economy that was not yet defunct. In short, the word "Darwinian" was often applied to that which was not. Yet, after a long road scattered with booby traps and detours, the scientific question gradually came to the fore. According to Y. Conry, the *Rapport sur les progrès de la zoologie* of 1863 recorded "the end of a period of radical opposition to Darwinism—at least in the scientific realm, for its ideologies, as is their nature, lasted beyond these dates; by 1900, they had still not completely stopped imposing their views."[8] Like the bird of Minerva, they fly away late in the evening. This explains the long succession of Protestant theologians, scientists, and philosophers from Germany (Rothe), England (Drummond), and France (A. Sabatier and F. Leenhardt), as well as Catholics from the abbé Boulay to Maximilian Begouën. These men did not necessarily proscribe Darwin; nor were they necessarily Lamarckians. But Lamarckian themes colored their thinking, and like phagocytes they devoured any surviving Darwinian concepts.

At the time of our first retrogression (the end of the nineteenth century and the turn of the century), Darwinism was sufficiently well established so that one could not go back to Lamarck without running into Darwin. But even before meeting him, cell theory, pathological anatomy, biogeography, paleontology, and botany had brought back essential Lamarckian themes, such as the primacy of function over organ. Occasionally, it also happened that Darwin was used to take Lamarck to task in the manner of Cuvier; or that Darwin was used to campaign for Lamarck; or that Lamarck was used to make Darwin acceptable. During these years, Lamarck was either a "rival or an associate," not a precursor.[9] Some rather eclectic minds thought they could synthesize the two theories. In this case, natural selection was subordinated to "the principles of the conditions of existence," to the "environments." Lamarckian factors were interpreted as expressions of either a radical mechanism or a teleological outlook, depending on whether the action of circumstances on organization or the serial design was taken as the reference point. Some passages from Albert Gaudry[10] sound strikingly like Lamarck; although this great paleontologist claimed he had read Darwin with impassioned admiration, his rejection of chance threw him back to Lamarck.

Two symmetrical but inverse examples clearly illustrate the

complexity of the situation in which those perspicacious minds proved unable either to deny Darwin's greatness or to overlook Lamarck's. The neo-Lamarckian Giard faced his contemporary Haeckel, who was allegedly more Darwinian than Darwin. These two scientists can be compared only with respect to their openness to the two currents. In all other respects, they are radically different. Giard, the founder to the laboratory of Wimereux, contributed to the experimental verification of the concepts of variation and adaptation. "My fondest wish, I would almost say my only driving passion, is to see the work undertaken in our area contribute to the spread of those admirable studies that Darwin, Vogt, Claparède, Kowalevski, and Haeckel have disseminated for the last twenty years among all nations in which science has made the most rapid progress, and of those doctrines that have effected in the biological sciences a revolution comparable to the one Newton brought about in astronomy."[11] In spite of the Wimereux results, which reinforced Darwin's position, Giard always granted the conditions of existence primacy over selection. His student, Paul Allez, would do likewise. These men put natural selection into a Lamarckian orbit.

Giard's introductory lecture on 1 December 1888 is characteristic in this regard. From the outset, it put Darwin and Lamarck on a par as "the creators of the modern theory of evolution."[12] Giard paid an ardent tribute to Lamarck for his concept of species, the role of use, and the inheritance of acquired characters. He corrected the caricatures and pointed out the false criticisms. But he also reproached Lamarck for having excluded the possibility of lost species, which had been easily integrated into Darwin's argument. But Giard seemingly did not notice Darwin's procedural break with Lamarck. For Darwin, selection is not the key concept of a perfect theoretical system but an event-dependent, albeit permanent, process. This is why selection can be joined without dissonance to those Lamarckian notions that, from one point of view, have more in common with natural history than with the biology introduced by the author of the *Philosophie zoologique*. In this edifice, selection can have only a limited, preservative, not a creative, role. Otherwise, the so-called laws of adaptation and the inheritance of acquired characters would no longer be dominant. In this, Giard concurs with Quatrefages, Lanessan, and all those for whom "environments" bring about creations. The fundamental task is to discover the mechanism used. Such is the position of these unprepossessing scientists. They wander through Darwinism and select what they think they need to enrich and complete Lamarckism, to which they return laden with booty.

If in contrast we examine Haeckel's position—not because the two men are similar, but because their positions are curiously symmetrical—we find the same desire to avoid prejudice. But Haeckel's encounter with Darwinism resulted in complete acceptance; his reading of Lamarck therefore has a very different character than that of Giard. A passage from the *History of Creation* (1868),[13] used as a preface to the 1907 edition of the *Philosophie zoologique,* places Lamarck on the same level with Goethe and Darwin: "To him will always belong the immortal glory of having for the first time worked out the Theory of Descent as an independent scientific theory of the first order, and as the philosophical foundation of the whole science of Biology."[14] The creator of monism was sensitive to the breadth of Lamarck's vision, developed, as R. Burkhardt has correctly pointed out, by a "naturalist-philosopher" at a time when teaching at the Museum was becoming increasingly specialized. What overjoyed him was that this grandiose perspective was mechanistic. "As you see, Lamarck's work is fully and strictly mechanistic, that is, mechanical"—an epic without a final cause. In this respect, his reading of Lamarck was analogous to that of the Giard school. Like the latter, Haeckel saw in Lamarck a precursor, a victim of Cuvier and clerical conservatism, a scientist who made organisms and even manifestations of the spirit emerge from physicochemical forces. He gave Lamarck the honor of having originated adaptation, but however highly he thought of the genealogical theory, he blamed him for not having seen natural selection. Of course, Haeckel considered selection to be the essential element that makes Darwinism superior to Lamarckism. Natural selection had subsumed Lamarckian principles: "The fact that Lamarck's wounderful intellectual feat met with scarcely any recognition arises partly from the immense length of the gigantic stride with which he had advanced beyond the next fifty years, partly from its defective empirical foundation, and from the somewhat one-sided character of his arguments." But as a philosopher, Haeckel admired his views: "They are indeed astonishingly bold, grand, and far-reaching"; in short, his work is "a complete and strictly monistic system of nature."[15] Haeckel's thought was analytical in his scientific works, but synthetic in the strange monistic mechanism that adopted the tone of an epic lyricism to describe the "becoming" of nature.[16] A doctrine that would take into account only the facts would not get very far. "Philosophical thought alone can make a science of it." It alone can undergird a powerful unification in which ontogeny reproduces phylogeny. "As one can see, embryological history and paleontological history, ontogeny and phylogeny, lead to the same result."[17] Such is the "confession of faith of a naturalist." The expression is significant. Following Buchner,[18] and at the same time as Spencer, Haeckel produced the scientific ideology of a philosopher and, at the same time, an ideology of the scientific—to adopt Canguilhem's categories. He rose up against the churches, but in a certain sense replaced one faith with another. If the Latin verb *religere* does indeed signify "to bind," his monism is a quasi-religion. Haeckel sensed in Lamarck this ardent imagination, this visionary side he himself possessed and applied in a strange way to scientism. This is what carried his Darwinism into a unitary thrust, of which Darwin could not approve; this is what drew him to Lamarck, in spite of his Darwinism. In spite of this long road from Darwin to Lamarck by reading history backward, he was a Darwinian and not a Lamarckian—universal mechanism was guaranteed by natural selection, but threatened by the design of the series. Scientific progress went hand in hand with progress in philosophical thought and society. This much is obvious from the little tract in which Haeckel assails Virchow for his reactionary position against academic freedom (1878).[19] All the religious, philosophical and political ideologies are unified by the idea of progress and the struggle against reaction, that is, by evolutionary ideology. This is the Weimar of Goethe against Berlin, the prince as a paradoxical defender of freedom against the Prussian bureaucrat, the old leader of progressivism grown conservative.

It would be a mistake, however, to think that Haeckel was extrapolating the struggle for existence into the social sphere. On the contrary, he pointed out the dangers involved in "brutally carrying scientific theories into the domain of applied politics." Progress consists in struggling against the animal condition; with the arrival of man, evolution has a new meaning. The work of this towering German intellect thus found in Lamarck a broad vision, likewise inspired by freedom and progress. In Darwin, he found the scientific instrument that guaranteed the rigor of his doctrine; selection, the key to biological mechanism, itself linked to physicochemistry and to the network of causality made possible the exclusion of teleology. A convinced Darwinian had returned to the Lamarckian source.

It would be preposterous to compare Giard, a good biologist and an orderly Cartesian French mind, to the colossal Haeckel. In Lille, Giard belonged to the laical, that is, anticlerical faction, but did not create his own ideology. He stood for a notion of democracy that he expressed in the preface to his edition of the *Discours* of Lamarck, who for him was the symbol of progress and a victim of reactionary trends. The address of 1888 has the same thrust, and reaches even beyond Lamarck to Buffon's difficulties with religious dogmatism. In Giard, as in Haeckel, evolutionism was the sign of the march forward and the liberation of humanity. Giard was Lamarckian; he retained of Darwin only what complemented Lamarck. Haeckel was Darwinian, and retained of Lamarck only what could breathe philosophical breadth into Darwinian theory. But this did not prevent the two scientists from displaying analogies as different in form as the wings of a bird and the paws of a mammal. From the perspective of a generous and open mind, Lamarck shines gloriously under the retrospective light that Darwinism projects onto him. In our opinion, it is this relation that carries the most fruitful current of evolutionary thought. This is why we have neglected the scientists who denatured Lamarck's thought, either by opposing it to Darwin on the premise of a finalism foreign to him or by reducing it to Darwin and pursuing Cuvier's prosecution. Instead we have considered two thinkers who did not adopt such poor tactics. After Y. Conry's study, it would be superfluous to take up all the examples of this interaction between Darwin and Lamarck. The essential point is that it is scientific work in botany, geography, and even physiology (via the "internal environment" of Bernardin de Saint-Pierre, which intersects with the fluids in a surprising and, to be sure, distant manner, although they first seem to be survivals from the eighteenth century) that provokes a neo-Lamarckism that is far from being static and repetitious.

The religious reaction is manifest throughout Protestant as well as Catholic thought; both weighed the spiritual consequences of evolution. Haeckel claimed that it was necessary to choose between creation and evolution, that there was no third way. The religious camp, on the contrary, tried to defend the synthesis of creation and evolution, creation through evolution; the expression "evolutionary creation" recurs frequently. The idea, if not the term itself, occurs in Lamarck as well as in Darwin, both of whom had emphatically denied that they were atheists! "During the last century and in the first years of our own, several theologians who were aware of scientific problems (especially questions raised by the evolution of species) tried to present either a vision of science compatible with the Christian faith or a theology compatible with science."[20] Rothe's *Ethik* was apparently the first attempt of the sort. Inspired by Hegelianism, it described a process by which the divine is made real "by the passage of matter to life, and of life to spirit." This perspective is too broad to be termed a return to Lamarckism. The geologist Leenhardt wrote about his teacher Rothe and produced an ideology misleadingly termed "evolutionary creationism." Even though Lamarck was not explicitly used in the text, the substance was Lamarckian: compatibility of creation with evolution, active presence of the Supreme Author. The repudiation of Darwinism as mechanistic left room for God as motor. It is not clear whether Leenhardt interpreted Lamarck as an advocate of mechanism or finalism. Anomalies were seen as manifestations of activities ordered by the environment, but also susceptible to being countered in the case of man. Y. Conry sees a certain neo-Lamarckism in Leenhardt. As for Drummond,[21] he based his thought on the unicity of a law that governs the world and makes it pass through the material "kingdom" via the living "kingdom" to the spiritual (third) "kingdom." Evolution is the spiritualization of the natural. Most interesting of all is Sabatier,[22] a theologian at heart, who wrote deliberately ideological-scientific works and tried to show, against Renouvier, that Genesis did not require fixism.[23] He founded a Christian evolutionism that retraced the epic of nature through successive stages, each one of which gave birth to a later and superior one before reaching maturity. All of these stages subsisted as mutual conditions, right up to the final stage of the moral being: "For the ascending march of creation to lead to an ever closer resemblance with God . . . it was necessary for this God to put in nature the power or the rudiment of His attributes." The essential attribute was freedom.

It is surprising that among the scientists who were also theologians, or at least very involved in religious thought, the Darwinian theory of evolution provoked an ideological clarification that gave rise to a world view (*Weltanschauung*). The second half of the nineteenth century was marked by a veritable burst of evolutionary ideologies, religious as well as irreligious. In Sabatier's writings, Darwinian theory fit very precisely into the parallelism of ontogeny and phylogeny, for example, with regard to transitional organic forms such as those that form an array extending from the fishes to the higher vertebrates. This Protestant scientist was resolutely evolutionary, but he nevertheless did not welcome all of the Darwinian factors: his critique fell within the bounds of scientific procedure. But like the other spiritualist transformists, he was related to Lamarck by his invocation of the presence of an ordering power in the design of nature. In summary, men who felt that their religious convictions were threatened found in Lamarck the antidote to a Darwinism they nevertheless admired. "Natural selection," and chance in particular, aroused their antagonisms because not only was teleology excluded but evil was thereby rooted in fundamental principles of creation and hence became irreducible. But at the same time they were sincerely frightened on the political plane by the fact that a world of brutal competition made aspirations to progress very unlikely. Although Sabatier did not quote Lamarck, their views of nature as a manifestation of the Supreme Author suggest a convergence of thought.

As a counterpart to the liberalism of the Protestants, let us turn to the current of Catholic thinkers. Y. Conry has analyzed the predominantly tactical ploys of the abbé Boulay, Giard's adversary in Lille, whose work fits into the context of Pope Leo XIII's policies, that is, a "rallying" to modern science parallel to the trends in modern politics. One quotation will suffice: "By 'rallying' (*rallier*) to the theory of evolution, the goal of some of us is to take a position within the enemy's fortress and to show that, in the end, it belongs to us. This undertaking is at every point comparable to the "rallying" to the Republic. It presents the same advantages, the same difficulties, and the same dangers."[24] Boulay's thought is not very lofty. It is nevertheless significant that some dreamed of annexing Darwinian science to Catholic orthodoxy after expurgating, of course, its most compromising principles. These were thought to hold not only a threat to religion but also a political peril; the people might develop a taste for struggle[25] at the expense of a conciliatory social Catholicism that planned to temper pauperism by mutualism and the *rapprochement* of classes. (It seems worthwhile to mention in passing this willingness to coopt a scientific result shown by a religious ideology linked to a political ideology, both of which adopt extrinsically motivated tactics.)

At another level, we shall consider an example that fits more appropriately in the category of speculation than of tactic; in this respect it is close to Sabatier. This is the case of a Catholic paleontologist, Count Maximilien Begouën, who wrote *La Création évolutive* and whose *La Vibration vitale* was published posthumously by his sons in 1885. One of the latter, also a Count Begouën and a "prehistorian" at the university of Toulouse, published a very interesting document on the transformist movement in Catholic circles around 1880.[26] Here one sees ideology in action, not within scientific procedure (as in Haeckel) or outside of it as a tactic of aggression and annexation (as in the abbé Boulay), but in an intellectual effort to synthesize science and Catholic orthodoxy in a sort of intellectual embrace and generalized vision. At a speculative level, Begouën betrays the desire to baptize transformism, to be sure. But he also testifies to a difficult struggle within Catholicism between supporters and opponents of evolution, a point that is often overlooked; and finally, remarkably, he also professes an enormous admiration for Darwin, who stimulates the explicit return to Lamarck, but implies theses that modify Darwin in a Lamarckian sense. From this testimony there emerges a treatment of Darwinism similar to that found later in Teilhard de Chardin, albeit in a more self-conscious form. This treatment of Darwinian science has the aim of discovering an "area of conciliation."[27] The author announces in his study that "the laws of the Darwinian school, very true to a certain extent, are only clear and logical for him who sees in them the instrument of divine law."[28] This law is made manifest by a slow and majestic unfolding of creation in time. Time is charged with this "creative power" of forming groups of beings, and progress in this time "necessarily leads to the idea of God." Begouën's reproach to Darwin is precisely that he allowed God into his first writings by giving Him a very restricted role. "Once creation had been completed, He no longer reappeared and his intervention could be felt nowhere." Thus the author's idea of science includes the allusion to, and the reminder of, God's action. The ideological and scientific levels are blatantly mixed. Darwin is blamed not for the negation of God, but for His absence from a work of biology. Scientific causality is not a self-sufficient system. Later Teilhard would adopt a different position by affirming that God is not revealed in the gaps of causality. As Begouën put it, "Transformism is not only 'completed' by the idea of God . . . but must be inhabited by God."[29] Elsewhere Begouën wrote, "Darwin pursued some of the most interesting research in natural history, and he brought to light many facts that had escaped notice or were known only to scientists. But carried away by the subtlety of his intelligence, and wishing perhaps to put forward a philosophical thesis, he formulated laws that are incomplete in that they attribute to matter an independence it does not possess."[30]

Count Begouën, who knew of Haeckel's work, evidently feared that Darwin would proceed from the absence of God to the autonomy of matter. The product of Darwin's subtlety that seemed so frightful to the Catholic scientist was surely the theoretical system founded on chance. Begouën welcomed the facts, but rejected the system in spite of his admiration for Darwin's work. Moreover, he criticized selection by focusing on the issue of species, whose variability he thought Darwin had exaggerated at the expense of their fixity. He conceded that the latter had presented a remarkable defence of evolution, but wanted to see the Superior Will appear Who "contributes to the execution of laws through all of the perturbing causes." His transition from Darwin to Haeckel proves that in Darwin he feared the possibility of a Haeckel.[31] It is interesting to note that Begouën ultimately returned to Lamarckian positions with regard to the place of God within a philosophical biology. Although he alluded to Lamarck at the beginning of the *Vibration vitale,* yet it was for Darwin that Begouën saved his admiring epithets.[32]

Count Begouën belonged to that group of scientists from Toulouse who applauded the valiant efforts of the Catholic supporters of transformism against the literalists and integralists. While the abbé Boulay was fighting non-Catholic Darwinians, the latter were fighting the reactionary forces within the Catholic Church, an aspect of the issue that deserves to be emphasized. The two Begouëns, father and son, entered the arena against the fanaticism of believers and nonbelievers alike. But whereas they denounced ideological pressure on questions that ought to have been treated objectively (e.g., the well-known question of the time dimension), they fell into a trap that was perhaps more serious: They brought religious elements into the scientific procedure itself. Along with their colleagues Cartailhac, Cartet, Garrigou, and Joly, they violently attacked the counts of Adhémar and Lucenzon and the Malefosse brothers. Beyond this heroic flourish in 1880, they worked in association with Quatrefages, Gaudry, Nadailhac, the astronomer Faye, Crookes, and others: "At the time, the reaction against Cuvier's theories was in full swing . . . and by the same stroke an excessive continuism was espoused. We were convinced that we were laying the groundwork for the victory of a spiritualist evolutionism; we were certain that materialism was powerless in the face of scientific proof, and intellectually bankrupt."[33] For them, scientific explanation could not stop at "secondary causes"; it had to go on to the first cause. This view of science is reminiscent of an era well before Lamarck. At the time, Nadailhac was publishing *Les Premiers Hommes*. Yet although these excellent scientists fought tooth and nail against the imperialism of narrow religious concepts, they were not opposed to the illumination of science by religion. Monsignor d'Hulst, a defender of transformism, would claim from the pulpit at Notre-Dame Cathedral, "With God at the origin of being, with God at the endpoint of progress, with God along the sides of the column to direct and support its movement, evolution is admissible."[34] Here history was retraced not merely to Lamarck but to Moses. "Like Darwin, Moses presents a progression among living beings."[35] "Well before Darwin, Moses had laid the foundations of a progressivist system. The work of the English scientist pays a brilliant, albeit involuntary, tribute to the work of Genesis."[36]

It is a curious stance, this fascinated and admiring consideration of Darwin, this return to Lamarck through a Darwin who in some cases, is enshrined as the dominant figure or, in others, merges with his predecessor. This movement backward underwent a revival in lay and religious circles of the early twentieth century, as well as among the biologists who studied living forms. During this time, thanks to genetics and molecular biology, neo-Darwinism emerged as sovereign and Lamarck faded away. Yves Delage, who was still strongly influenced by Weismann, has given an impartial account of the debate between neo-Lamarckians and neo-Darwinians.[37] He thought that no critical experiment had yet decided between the innate and the acquired; he did not settle the issue in favor of one camp or the other. In fact, Lamarck's name tended to become a mere reference point, in particular to show that the transmission of acquired characters, the basis of Lamarckism, was still upheld by the neo-Lamarckians, but completely rejected by the neo-Darwinians, in contrast to Darwin himself. When L. Cuénot rejected chance around 1925[38] and again shortly before his death in *L'Évolution biologique,* he did not refer to Lamarck either, not even when he discussed adaptation. But the examples he used were often Lamarckian; the facts of adaptation were simultaneously facts of finalization: the reference to Lamarck was implicitly present between the lines, although it was no longer customary to adopt the title Lamarckian. Objections were raised against "universal chance," except among the neo-Darwinians; "antichance" was invoked. Appeals to the Supreme Author's design disappeared. In spite of the Catholic paleontologists and prehistorians in Toulouse, God no longer illuminated science from within.

One philosopher, who was neither Catholic nor Protestant but unquestionably an idealist and a spiritualist, sums up this long post-Darwinian period. He returned explicitly to Lamarck for the justification of his own evolutionary philosophy. In 1907, Bergson published *L'Évolution créatrice,* which like his earlier books represented an "exploitation"—to use Althusser's terminology—of science: "Of all the present forms of evolutionism, neo-Lamarckism is the only one that is capable of accommodating an internal psychological principle of development, although its does not necessarily appeal to it."[39] Bergson's philosophy is a philosophy of duration, as he himself carefully pointed out. This duration, whose essense is psychological, takes the form of a "vital impulse" (*élan vital*) at the biological level. Clearly, the privileged biological tendency will be the one that leads to the Bergsonian understanding of duration. On the basis of his earlier work, Bergson believed that the psychological model could be transferred to every other level. This cannot be held against him. What he sought in Lamarck, through a severe critique of Darwin, was an evolutionism founded on this internal principle. In reality, he went back to the Lamarck popularized by Cuvier, the Lamarck in which will substitutes for need, the Lamarck that Darwin encountered at the University of Edinburgh. In the Lamarckian concept of effort, capable of creating an organ through the exercise of a function, Bergson found an analogue to the creative power of his "vital impulse." To the mechanistic Lamarck, whom he knew and quoted,[40] Bergson preferred the Lamarck of the will favored by Cope. As Bergson stated it, the word "effort" then need only be taken "in a more profound, more psychological sense than any neo-Lamarckian supposes,"[41] and the result is Bergson's metaphysics. This did not prevent Bergson from recognizing the merits of Darwin and the neo-Darwinians. But he did reject the account of the formation of organs by the accumulation of imperceptible variations. His famous critique of Darwin's account of the formation of the eye in the vertebrates epitomized for him the inadequacy of Darwinian factors. In any case, the source of life could not be chance. Life is a creative evolution beginning from a center of "springing forth" (*jaillissement*): "When I speak of a center from which worlds shoot out like rockets in a fireworks display—provided, however, that I do not present this center as a *thing*, but as a continuity of shooting out—God, thus defined, has nothing of the already made; He is unceasing life, action, freedom."[42]

In the transition from this cosmic God, who eventually drew lightning from the Index, to the God of religions,[43] Bergson showed that the "vital impulse"—an impulse with monistic or pantheistic traits—was only an inverted teleology, and that his philosophy was religious. Thus the philosophy of *L'Évolution créatrice* presented a Lamarckism that was a deforming ideology of Lamarckian theory. In a 1915 text on French philosophy, for which Bergson selected thinkers and excerpts that pointed to what he termed "the author of creative evolution," he wrote, "It is only now that Lamarck is being given his due. This naturalist, who was also a philosopher, is the true author of biological evolutionism. . . ." After having praised Darwin for "sticking to the facts," for having discovered the role of competition in selection, Bergson restricted these factors to a preservative role, as the religious thinkers had done; and like them, he blamed Darwin for not having given an account of causes, whereas Lamarck had attempted it. He concluded that "more than one naturalist today is returning to Lamarck, either to combine Lamarckism with Darwinism or to replace Darwinism with a perfected Lamarckism."[44] In 1907, and later in 1915, a French philosopher had thus explicitly returned to Lamarck by what he called a "transfiguration" of science—what the Althusserian school would call a theoretical "exploitation" of evolution.

Bergson never revised this view. And yet, while he was justifying Lamarck, while a certain Lamarckian current persisted in the United States and in France (e.g., Wintrebert), the triumphal successes of genetics and molecular biology and the discovery of DNA were bringing back into the limelight the thesis of the innateness of acquired characters and the factors of competition, selection, and chance.

Where does evolutionary biology stand today? What has happened to neo-Lamarckism and neo-Darwinism? Are questions still framed in terms of these two camps? Certainly it is still possible to point to neo-Darwinians and to neo-Lamarckians—even to Lysenkists, and not in the USSR. A look at the problem of innate versus acquired characters shows these partisans locked in battle. But we have also noted that while Jacques Monod was triumphantly stating positions that he himself termed Cartesian because mechanistic, researchers were exposing the action of external circumstances upon the genes, of RNA on DNA, and so forth. Precisely when selective action, chance, and the exclusive message from the genotype to the phenotype seemed to be incontestable, research was raising questions about Darwinism. "Evolutionism evolves," wrote A. Jacquard in September 1976.[45] R. Lewontin, who was studying the relation between evolution and genetics, considered selection to be incompatible with genetic polymorphism and the genes to be neutral; chance remained.[46] As Nei presents it, population genetics at the molecular level gives primacy to mutation over selection. Modifications would therefore be genotypical, not phenotypical. As for chance, A. Jacquard in a March 1975 article finds its meaning difficult to elucidate. Is it indeterminism, unexplained determinism, a reality too complex to be known that leads to partial information? Jacquard's recent book[47] shows the extent to which population genetics is central to the formulation of these questions, which lead us back to the prehistory of this field in Darwin. E. O. Wilson's sociobiology integrates neo-Darwinism rather well: The fitness of a gene in competition and selection can be measured by its frequency in populations.

On his account, animal behavior would not only be regressive but would lead to cooperation. Wilson calls these behaviors altruistic. The results of these studies have not met with unanimous approval, but they do stimulate thinking on the issue. At the level of these structures and groups, individuals appear flexible and modifiable. One wonders, however, if this is not an instance of extrapolation from group psychosociology.

To be sure, all of this gives the impression of a smorgasbord of examples—superficial, panoramic, lacking in rigor. But on the one hand, we are scarcely in a position to evaluate this work in progress. On the other, the confrontations and controversies have the character of a Babel; each discipline speaks its own language of molecules, Mendelian characters, ethology, biosociology, and so forth. Neo-Darwinism is in crisis; the tendency now is to integrate it into more general theories. But neo-Lamarckism is also in crisis. A new problem-set should be free to emerge from prior ideological involvements, but it seems that its factors cannot avoid recapture.

In today's perspective, when biological research is proceeding at an accelerated pace, yesterday looks like ancient history—a sometimes prestigious history. I am thinking in particular of Jacques Monod's *Leçon inaugurale* in 1967. This highly suggestive piece was a sort of retrospective of the mechanism he defended and the teleology he opposed. Most surprising was the fact that in his discourse he was seemingly unaware that what he opposed was vitiating what he proposed to defend. It was difficult to understand how this Nobel laureate, who appeared among others as the purveyor of a physical theory of heredity (the molecular theory of the genetic code), that this Nobel laureate for 1962 wanted to communicate his "natural philosophy," like Lamarck in 1809. It was difficult to understand why the terms of the debate still involved teleology, as they had at the time of Bergson, L. Cuénot, the nineteenth century idealists, Cuvier, and Lamarck. It was difficult to understand why Monod took as his chief opponent Teilhard de Chardin, a thinker who, with slightly different materials, reflects in our own time the reaction of nineteenth-century religious thinkers to evolution. Were Teilhard's views so contagious that Monod himself had to use them, albeit in a strange way? Here was a scientist who attached himself to Darwin "formally," while pointing out that the latter of course could not have had in his own time "any idea of the chemical mechanisms of reproductive invariance" and their perturbations, "that the selective theory of evolution" derives all of its significance from the latest discoveries. Here was a scientist who denied that "*invariance* is *safeguarded, ontogeny guided, evolution oriented* by an initial telenomic principle . . . ."[48] The problem was not that he refused to see the telenomy that characterizes living beings, but rather that he located it under the dependency of "emergence." The use of this term always introduces an ambiguity, which is aggravated by the definition of emergence in terms of "evolutionary creation" and "increasing complexity."[49] Jacques Monod's definition uses the very terms that from Rothe to Sabatier, Begouën, and Teilhard always reappear in the spiritualist attitude. It would have been much simpler to stick to the emergence in DNA. Moreover, the term almost completely disappeared from *Chance and Necessity* following Althusser's critique in his 1967 course for scientists. The vocabulary of the *Leçon* included terms such as "biosphere" and "noosphere," both of them clearly Teilhardian, and the second strikingly idealist. By returning to themes we discussed elsewhere,[50] we wanted to conclude this chapter by showing that history in reverse is not only a succession of vague retrospectives, each of which is born from the preceding; it is especially the weight of a past that does not stop obsessing scientists and provokes a return that is tirelessly repeated. Jacques Monod the neo-Darwinian dreamed of freeing men from ideologies of consolation and salvation. In so doing, he has based science exclusive of values on research—a discipline that rests on a choice of values, as he himself both knows and says. The man who suggests himself as a disciple of a certain brand of nihilism is probably a metaphysician without knowing it. (Or perhaps he does know it.) Sisyphus as philosopher is constantly rebuilding a vision that in every historical period rolls back down the slope.

And now the world of scientists and thinkers is facing the ruins of its constructions. Mankind hesitates between the mechanistic universe of molecules and the history of visible forms; the first might lead to a stark dualism of objective reality and values that we construct at our own peril; the second, by considering the history of living beings, extrapolates the achievement of history and gives it meaning and transcendence. Monod the biochemist expresses the first attitude; Teilhard the paleontologist displays in an almost naive way, albeit with the power of a seer, a metaevolutionary eschatology.

The mirage of the past does not stop at Lamarck or Cuvier; it loses itself in a very ancient idea, according to which, in Democritus's words "all that exists in the universe is the fruit of chance and necessity,"[51] whereas the religions promise us sunlight as we leave the cave.

Conclusion: Lamarck in the Past, Present, and Future

Lamarck has been the subject of verbs indiscriminately conjugated in the past, present, and future tenses. Sometimes, the contradiction is blatant, for example, when a commentator or a biographer calls him a man of the eighteenth century and a prophet of the late nineteenth or early twentieth centuries. Between these two dimensions of time, the present collapses into nothingness. The design of nature and its author, the linear series, the transmission of acquired characters, and perhaps even the primacy of function over the organ are discussed in the past; his vision of transformism, descendance, evolution, and real adaptation in the future. And in the present, Lamarck is divided between his competence in the study of the invertebrates on the one hand and his odd divagations, his conjectures about all the sciences, the philosophy of nature, and the transformation of species on the other. The giraffe stretches its neck to reach the branches; the shore bird stretches its legs to avoid getting wet. The neck becomes elongated; the bird grows taller. A weird scientist is taking a walk under the arbors of the *Jardin des Plantes* while he ruminates his extravagant ideas. Within the walls of the long buildings where vast collections rest, he tirelessly observes and classifies. Time, like a wave, seems to engulf him, to make him surface again, less suddenly, to be sure, than a sketchy history suggests. Such is the *cliché* of Jean-Baptiste Pierre-Antoine de Monet, chevalier de Lamarck, alias—in revolutionary language—Citizen Lamarck.

The present study has certainly not been a painstaking investigation of Lamarck's papers or a patient and complete reconstitution of all that he wrote. Richard W. Burkhardt, professor of the history and philosophy of science at the University of Illinois, has recently done this in the United States, where, since Cope, Packard, and many others, an uninterrupted Lamarckian tradition lived on. Instead, we have attempted an analysis of the key concepts of Lamarckian theory: nature; the genealogy of groups manifested in the "series" modulated by "circumstances"; the transmission of acquired characters; use and disuse, which anticipate and introduce adaptations. These interrelated concepts form a system, as Burkhardt saw clearly when he entitled his book *The Spirit of System*. But Lamarck did not stop with the conceptual system that structures his transformationist biology. He also wanted to systematize the sciences in relation to one another. Against the stream of specialization that was already obvious in the assignment of teaching chairs in his own time, he wanted to be what the Orientals call a "general thinker," a naturalist-philosopher, so much so that Bergson includes him among the philosophers. We did not try to avoid the Darwin-Lamarck confrontation; we approached it through the analysis of concepts, especially that of adaptation. We examined it through the waves of thinkers and scientists who, after the *Origin of Species*, brought Lamarck back into question and into the limelight in relation to Darwin.

What are the results of this work? Can Lamarck be the subject of a present-tense conjugation? What relation does this present bear to that past and future whose "presence" we discerned in his work? Does the term "precursor" have any meaning beyond the myth? If so, does the term help identify the exact place where Lamarck stands? And is this place a peak from which lead roads that must be trod if we are to understand contemporary biology? These are the problems a conclusion must be in a position to formulate, if not to solve, more adequately than a preface.

To wonder in which dimension of time such or such a use of concepts is found does not consist in forgetting precisely that they are in use, and therefore taking their content as fixed. It is then too easy to speak of the idea of nature in the past, of adaptation in the future, and so forth. This is simply another way of transferring to science that brand of "source positivism" practiced for a time in philosophy and literature. This pseudomethod destroys the new by decomposing it into ancient elements. In contrast, we have tried to show how nature, the series, and the transmission of acquired characters functioned in Lamarck's own perspective. Which is to say that our own set of questions should not be predetermined by the neo-Lamarckians, insofar as they asked of Lamarck questions that Darwin had asked in order to get Lamarckian answers they could turn against Darwin's. Lamarck never asked these questions. Each of the preceding chapters has dealt with one concept, which we tried to analyze in the context of Lamarck's time. When we confronted Lamarckian "adaptation" with the Darwinian concept, the point was to bring out the difference between these two periods. And when we examined the opposition of two currents, it was always with the same objective: to reconstruct the structures of Lamarck's period. Burkhardt has described these environments with a remarkable wealth of documentation in order to emphasize "the importance of the structural transformation in scientific life at the end of the eighteenth century in France."[1]

Our study of the idea of nature did not dwell on the writings of Rousseau or others; such was not our aim. It was to show that the concept of nature in Lamarck allows one to see the transition from a power thoroughly penetrated with divine action to a universe of movement governed by necessary laws. The textual variations do not lead to a complete eviction of ideology (particularly because of the idea of "series"), but nature fills a theoretical function since she takes charge of the execution of the divine plan. The delegation of the supernatural to the natural, supported by a juridical-political scheme, forces the theological to retreat. This is what Lamarck did in his present by using a traditional idea as a tool. The idea of series, which we then analyzed as an ordering of living beings in nature, further accentuates the presentness of eighteenth-century authors. In the *Histoire naturelle des animaux sans vertèbres*, Lamarck himself was unable to retain a series with a single trunk and had to double it. Does the latter concept not have a theoretical function? In my opinion, it was through the idea of series that Lamarck perceived both the constancy and the variability of living beings and the limitations of the species. He then searched for a natural order at the level of large groups or masses, which he classified by ordering them from the most rudimentary to the most complex, after having considered them in inverse order, beginning with mankind as a criterion. It was in moving from logical succession to the succession of chronological production that evolutionism impressed itself upon him. R. Burkhardt thinks that Lamarck was translating Linnaeus's natural economy into a genealogical transformism. But he also notes that Lamarck never said this. On the contrary, the texts in which he discussed transformation (especially his *Discours*) are very often those in which he raised the problem of species—the transition from variety to species and the variability of species. The spectacle of breeders and of nature did the rest.[2]

This long summary was intended to show that the idea of series, like that of nature, drew on the cultural legacy of the past, but became a tool for the construction of a new system of thought. This is also the case for the idea of circumstance, which in my opinion was a consequence of the linear idea of series. More specifically, it is a consequence of the contradiction between this model and the ramifications and diversifications that do not conform to it and therefore require an explanation. The notion of transformation then intervenes at two different levels:

that of the divine plan and its complexification; and that caused by the pressure of circumstances and the organism's response. This is the locus of adaptation. Lamarck clearly did not use the term; nor did he belong to the natural theology tradition—either he did not want to or he never considered it. The first suppostion seems to be the right one. He was here refusing a construction block from the past that he had to refrain from drawing out of the "natural harmonies." He therefore introduced the influence of use, of the primacy of function over the organ. There is no theological residue here, but there is an insufficiency of proof. There is also an argument aimed directly at fixism, and Cuvier perceived it as such. It should be noted that in the chain of Lamarck's ideas, the concept of use is the one Darwin would adopt by integrating it into his own theoretical system. In the same way, the notion of transmission of acquired characters—borrowed from the past and somewhat mythical, although not metaphysical—will be taken up by the Darwinian system. In Lamarck, the latter concept undergirds the effect of use and disuse, which without it would not go beyond the individual or cause transformism.

In short, the entire chain of Lamarckian ideas is borrowed from the past or observation. But the whole of it functions within a program for establishing a biology liberated from confessional implications. The materials are old, anchored in the past. The program is squarely situated in Lamarck's own times, if one concedes that a historical period is not a finished entity, but something in process by virtue of the activities of certain minds or human groups. The chief traits of this present-in-process are the specificity of biology—a discovery made simultaneously at several locations throughout Europe—and the transformation of species within a genealogical perspective. Lamarck's struggle to get his ideas admitted does not allow his duel with Cuvier to be treated as a legend, even if the latter's comparative anatomy would eventually do more for evolution than Lamarck's broad, unproved perspectives.

We now come to the final point: Is Lamarck that giant who posthumously leaps over fifty years of silence to come to life again as a precursor? Did he foretell the future without being heard? At the beginning of this book, we adopted Canguilhem's noteworthy views on the concept of precursor. We are now in a position to add some nuances to the expression "myth of the precursor."

If a precursor is one who sees and demonstrates a stage of science that will later be repeated and then heard, then the notion of precursor is a romantic myth. Even Mendel was not a precursor.[3] After deliberately concluding that a usable answer does not exist, Canguilhem's answer turns to metaphor: "Should we rest content with an image, and say that Mendel's work was like a child born prematurely, who is left to die because no one is prepared to receive it?"[4] Lamarck's case is quite different; he did not "run ahead of all of his contemporaries." The best of them, like Cuvier and Geoffroy Saint-Hilaire, were running beside him on their own track. He did not stop "on the track where others, after him, run to the finish line,"[5] for the connection between Lamarck and Darwin is not the linear relation of a relay run. Lamarck built one system of evolution; Darwin built an entirely different one. One crumbled; the other stood. Lamarck's relation to the future is not that of a precursor, but that of a silent witness (*présence*) to later developments, and especially in our own time. In *Le Monde* of 5–6 February 1978, Pierre Viansson-Ponté opens an important article with the following words: "Heredity or the environment, innate or acquired characters? This is a great debate, perhaps even *the* great debate." The Darwinian environment is not a Lamarckian concept, circumstances partly fill the role of the environment, and environments are part of circumstances. Lamarck illustrated the acquired. We know that it was one of the unquestioned bases of his thought. Viansson-Ponté adds; "The controversy is not a new one; it has been under way for more than a century. If it is on the rebound today, it is because developments in the life sciences have now given the partisans of each thesis a host of new arguments with which they passionately, and sometimes furiously, assail their opponents." And this furor engulfs not only biology but also reflections on "the origin of inequality in man" and the very legitimate ideologies for which people live and die.

It is therefore more appropriate, in my opinion, to refer to Lamarck's relevance (*présence*) to today's world, to a return to Lamarck in order to illuminate such or such contemporary question. For the effort to return to Lamarck implies precisely that one cannot transfer him as he is into our society in order finally to understand his prophetic discourse. Let us therefore make an effort to situate him in his own environment by attempting to discern the conjunctions of ideology and science that occur in his thought. But in so doing, we cannot forget that this period, this thought, the scientific and ideological elements that mingle in it are all reconstructed from our own set of questions. We offer the reader a Lamarck seen in his own time—within the dimensions of a perspective produced in the contemporary world. In this world preoccupied with the concern to understand the role of ideologies, Lamarck appears as a scientist involved in the perilous exercise that requires, as Canguilhem has put it, "a certain precedence of intellectual adventure over rationalization." But is not rationalization present in this adventure, and does not the adventure remain at the heart of rationalization?

Notes

Foreword

1 See his article in *Le Monde* of 5–6 February 1978, which summarizes the discussions of the issue published in the same newspaper throughout 1977.

Preface

1 Ernst Haeckel, *The History of Creation: Or the Development of the Earth and its Inhabitants by the Action of Material Causes* 2 Volumes, transl. E. R. Lankester (New York: D. Appleton and Co., 1883), I, p. 111.

2 Edmond Perrier, *Lamarck et le transformisme actuel* (Paris: Imprimerie Nationale, 1893), p. 5.

3 Georges Canguilhem, *Études d'histoire et de philosophie des sciences* (Paris: Vrin, 1970), introduction, pp. 20ff.

4 Georges Canguilhem, *Idéologie et rationalité* (Paris: Vrin, 1977), introduction, p. 13.

5 *Ibid.*, p. 109.

6 G.Canguilhem, *Études,* p. 21.

7 François Jacob, *The Logic of Life,* transl. B. E. Spillmann (New York: Vintage Books, 1976), p. 10.

8 G. Canguilhem, *Idéologie et rationalité,* p. 13.

9 *Idem.*

10 F. Jacob, *Logic of Life,* pp. 10–11.

11 G. Canguilhem, *Etudes,* p. 22.

12 Alfred Giard, *Controverses transformistes* (Paris: Naud, 1903), p. 17.

13 *Ibid.*, p. 18.

14 E. Haeckel, *History of Creation,* I, pp. 111–112.

15 Lamarck, *Oeuvres choisies* (Paris: Flammarion [1913?]), introduction by F. Le Dantec, p. 2.

16 *Ibid.*, pp. 2–3.

17 E. Perrier, *Lamarck,* p. 12.

18 Alphaeus Packard, *Lamarck, the Founder of Evolution. His Life and Work* (New York: Longmans, Green and Co., 1901), p. 76.

Chapter 1

1 A. S. Packard, *Lamarck, the Founder of Evolution. His Life and Work* (New York: Longmans, Green and Co., 1901), p. vi. Since Packard's work, several Lamarck manuscripts have been uncovered; see Morton Wheeler and Thomas Barbours, eds., *The Lamarck Manuscripts at Harvard* (Cambridge, MA: Harvard University Press, 1933); Max Vachon, Georges Rousseau and Yves Laissus, eds., *Inédits de Lamarck, d'après les manuscrits conservés à la bibliothèque centrale du Museum d'Histoire Naturelle de Paris* (Paris: Masson, 1972); Pierre-Paul Grassé, "La biologie, text inédit de Lamarck," *Revue scientifique* 5 (1944), 267–276.

2 The house was sold to the de Guilbon family in 1870. It was destroyed by the German invasion in 1914.

3 See E. Perrier, *Lamarck et le transformisme actuel* (Paris: Imprimerie Nationale, 1893), p. 13.

4 Marcel Landrieu deserves credit for the discovery; see his *Lamarck, le fondateur du transformisme. Sa vie, son oeuvre* (Paris: Au siège de la société zoologique de France, 1909).

5 Académie des Sciences, Paris, Scientific manuscripts of Cuvier, carton # 156, especially the "lettres des enfants de Lamarck."

6 The *Revue de Gascogne* for 1876, and the *Revue biographique de la Société malacologique de France* for 1886.

7 See Landrieu, *Lamarck,* p. 471.

8 The date is 1760 for Edmond Perrier and Alphaeus Packard, but 1761 for Landrieu. The latter mistakenly attributes Lamarck's departure to the expulsion of the Jesuits from Amiens, which took place only in 1764.

9 See E. Perrier, *Lamarck.*

10 Buffon, *Histoire naturelle,* IV (Paris, 1755) p. 59; quoted by A. Giard, *Controverses transformistes* (Paris: Naud, 1904), p. 11.

11 The *Dictionnaire* was published from 1782 on.

12 There is also a *rue Lamarck* in Paris, but not in the immediate vicinity of the *Jardin des Plantes.*

13 E. Perrier, *Lamarck,* p. 9.

14 Louis Blanc, *Histoire de la Révolution française,* 12 volumes (Paris, 1847–1862), volume 9, book X, chapter xii.

15 *Ibid.,* p. 412.

16 Hérault de Séchelles, *Voyage à Montbard,* quoted by Giard, *Controverses,* p. 11, note 1.

17 Lamarck, *Système analytique des connaisances positives de l'homme* (Paris, 1820), p. 90.

18 See Bernard Mantoy, *Jean-Baptiste de Lamarck* (Paris: Seghers, 1968). Based on information provided by Mr. Laissus, the director of the central library of the Museum, Mantoy thinks that even the very late writings are in Lamarck's hand.

19 See Louis Althusser, *Philosophie*

et philosophie spontanée des savants (Paris: Maspero, 1967), and chapter 3.

20 See chapter 6.

21 See the appended bibliography and more generally the references to Lamarck's books, whose titles are mentioned here only to support reflections about the general development of his thought.

22 Lamarck, *Système des animaux sans vertèbres* (Paris, 1801).

23 Lamarck, *Philosophie zoologique* (Paris, 1809).

24 Lamarck, *Histoire naturelle des animaux sans vertèbres,* 7 volumes (Paris, 1815–1822), introduction.

25 Lamarck, *Système analytique des connaissances positives de l'homme* (Paris, 1820).

26 See chapter 3 for a fuller discussion of this Lamarckian concept.

27 Hans Reiff and Cornelius Schuurmans, "La prévision météorologique," *La Recherche*, February 1976. Interestingly, in Lamarck's *Hydrogéologie* air is treated as a fluid.

28 But not the imaginary.

29 Thomas Henry Huxley, "Evolution in Biology" in *Darwiniana. Essays by T. H. Huxley* (New York: Appleton and Co., 1896), p. 212.

30 On this issue, Buffon is an illustrious exception.

31 B. Mantoy, *Lamarck,* p. 56.

32 In his book *Flaubert,* Sartre emphasized this point.

33 I. Kant's *Critique of Judgment* is one of the exceptions.

Chapter 2

1 *Conversations of Goethe with Eckermann and Soret* (London: George Bell and Sons, 1875), p. 131, hereafter cited as Goethe, *Conversations.*

2 *Ibid.,* p. 480.

3 See E. Haeckel, *History of Creation* (New York, 1883), lesson 2.

4 See F. Jacob, *Logic of Life* (New York, 1976), chapter 2.

5 See the beginning of the *Discours de l'an X,* in A. Giard, ed., *Les Discours d'ouverture...par J. B. Lamarck* (Paris, 1907), hereafter cited as Lamarck, *Discours.*

6 Lamarck, *Discours de mai 1806,* beginning, in *Discours.*

7 *Ibid.,* pp. 113, 115.

8 The word "biology" appears in the *Discours de l'an XI* and in the foreward to the *Philosophie zoologique.*

9 Lamarck, *Histoire naturelle des animaux sans vertèbres,* introduction, part 6, p. 36.

10 See the use of this word in Lamarck, *Histoire naturelle,* p. 41.

11 Lamarck, *Histoire naturelle,* introduction, p. 36.

12 *Ibid.,* p. 38.

13 The administration that the French Revolution put into place replaced the divine right of kingship.

14 Lamarck, *Histoire naturelle,* introduction, p. 41.

15 *Ibid.*, p. 42.

16 See the introduction to Jacob's *Logic of Life.*

17 Lamarck, *Histoire naturelle,* introduction, p. 43.

18 *Ibid.*, p. 48.

19 *Ibid.*, p. 47.

20 *Ibid.*, pp. 47–48.

21 *Ibid.*, pp. 45–46.

22 *Idem.*

23 *Ibid.*, p. 51.

24 *Ibid.*, p. 54.

25 The *Harmonies* followed the *Études de la nature* (1784–1790). Lamarck was undoubtedly acquainted with Bernardin de Saint-Pierre, who was superintendent of the *Jardin du Roi.*

26 Jacques Monod,*Chance and Necessity,* transl. A. Wainhouse (New York: Vintage Books, 1971), p. 21.

27 F. Jacob, *Logic of Life,* p. 2.

28 Lamarck, *Histoire naturelle,* introduction, pp. 38, 40–41, 43, 47.

29 *Ibid.*, p. 57.

30 *Ibid.*, p. 53.

31 *Ibid.*, pp. 62ff.

32 *Ibid.*, pp. 63–64.

33 *Ibid.*, p. 64.

34 *Ibid.*, p. 66.

35 See the *Corpus général des philosophes français,* volume 45, pp. 153–154, 213–214 on sensation and Condillac; part 1, pp. 111–112, 142–143 on the origin of ideas; part 2, pp. 313–315 on observation.

36 See the beginning of the Preface to the first edition of Kant's *Critique of Pure Reason.*

37 *Corpus général des philosophes français,* volume 45, part 2, p. 270.

38 *Ibid.*, part 1, p. 91.

39 Yet Buffon had already suggested "subtle experiments."

40 Lamarck, *Discours,* p. 105.

41 *Ibid.*, p. 100.

42 *Ibid.*, pp. 100–101.

43 G. Canguilhem, Epigraph to the *Quatrième cahier pour l'analyse* (published by the circle for epistemology of the *Ecole normale supérieure,* September/October 1966).

44 G. Cuvier, *Discours sur les révolutions du globe,* p. 58, quoted in *Science contemporaine,* vol. I, XIXe siècle, p. 569.

45 See Schelling, *Ideen zu einer Philosophie der Natur,* French translation in F. W. J. von Schelling, *Essais* (Paris: Aubier, 1946), p. 68.

46 *Ibid.*, p. 96.

47 Xavier Léon, *Fichte et son temps,* 2 volumes in 3 (Paris: A. Colin, 1922–1927), II, p. 338.

48 His discovery of the intermaxillary bone in man, the convergence of some of his views with those of Geoffroy

Saint-Hilaire on the bone structure of the cranium, his plant studies, and many others are examples of his active interest in science.

49 Goethe, *Conversations,* p. 480.

50 *Ibid.,* p. 131.

51 *Ibid.,* p. 282.

52 He is here discussing warblers feeding little linnets.

53 *Ibid.,* p. 301.

54 *Ibid.,* p. 524.

55 *Ibid.,* p. 423[?].

Chapter 3
1 1907 ed., part 1, chapter 6, p. 105.

2 *Ibid.,* pp.107, 109.

3 See F. Jacob, *The Logic of Life* (New York: Vintage Books, 1976), chapter 2, "Organization."

4 Lamarck, *Philosophie zoologique,* part 1, chapter 5, p. 103.

5 G. Cuvier, *Le Règne animal* (1817). These "plans" are in fact *models.*

6 Alfred Giard, ed., *Discours d'ouverture...par J. B. Lamarck* (Paris, 1907; hereafter Lamarck, *Discours*), p. 50.

7 *Ibid.,* p. 86.

8 *Idem.*

9 *Ibid.,* p. 67.

10 *Ibid.,* p. 105.

11 Jacob, *Logic,* p. 74.

12 Lamarck, *Discours,* p. 119.

13 Jacob, *Logic,* p. 77.

14 *Ibid.,* p. 82.

15 *Ibid.,* pp. 100–101, especially p. 101: "Homology describes the correspondence of structures, analogy that of functions."

16 See chapter 5.

17 Lamarck, *Discours,* p. 91.

18 Even though Lamarck occasionally uses the term *milieu* in the singular, for example in the *Recherches* of 1802.

19 See chapter 2.

20 Lamarck, *Discours,* pp. 100–101.

21 *Ibid.,* p. 113.

22 *Ibid.,* p. 115.

23 See Jacob, *Logic,* chapter 3, especially pp. 142ff.

24 *Ibid.,* p. 130.

25 *Ibid.,* pp. 133–134.

26 *Ibid.,* p. 143.

27 This is the case even though the axial line of the Lamarckian series gives rise to ramifications.

28 Althusser's shorthand for "spontaneous philosophy of the scientist"; see note 51. This reflection is indebted to the questions posed by Althusser.

29 Henri Daudin, *Les Classes zoologiques et l'Idée de série animale en France à l'époque de Lamarck et*

Cuvier (1790–1830) (Paris: Alcan, 1926).

30 The soul, life principle and first entelechy of the body, includes a generic diversity since it is itself analyzed into nutritive, sensitive, and intellective. For Aristotle, the realm of beings endowed with a soul stands in opposition to the inorganic realm, even though gradations make the distinction difficult. See Aristotle, *On the Soul,* book II, 414–424b.

31 Aristotle, *History of the Animals,* book VIII, chapter 1, 588b, lines 4–6; translated by D'Arcy Thompson, in W. D. Ross and J. A. Smith, eds., *The Works of Aristotle,* IV (Oxford: Clarendon Press, 1967).

32 See among other Leibniz texts, *On Nature in Itself, or on the Force Residing in Created Things, and Their Actions* (1698), in Philip Wiener, ed., *Leibniz: Selections* (New York: Scribners, 1951), pp. 137–156.

33 Foucault claims that Charles Bonnet (*Oeuvres complètes,* III, p. 173) quoted a letter from Leibniz to Hermann on the chain of beings; see *Les Mots et les Choses* (Paris: Gallimard, 1966), p. 165, note 1.

34 H. Daudin, *Les Classes zoologiques,* pp. 196–197.

35 See M. Foucault, *Les Mots et les Choses,* pp. 164–165.

36 Title of a work by Charles Bonnet.

37 M. Foucault, *Les Mots et les Choses,* p. 165.

38 Anagram of "de Maillet."

39 Charles Bonnet, *La Palingénésie philosophique,* in *Oeuvres complètes* (Neuchatel: Fauche, 1779–1783), volumes 15 and 16, part 6, p. 252.

40 Charles Bonnet, *La Contemplation de la nature,* in Bonnet, *Oeuvres complètes,* volumes 8 and 9; chapter 17, p. 359.

41 *Ibid.,* part 2, chapter 10, p. 35.

42 *Idem.*

43 *Ibid.,* chapter 17, p. 366.

44 These problems are too complex to be treated here; they are the subject of chapter 4.

45 Henri Bergson, *Les deux sources de la morale et de la religion* (Paris: Alcan, 1932), chapter 3.

46 F. Jacob, *Logic,* p. 148.

47 Canguilhem, *Idéologie et rationalité* (Paris: Vrin, 1977), Introduction and part I (lecture delivered before the Institute of the History of Science and Technology of the Polish Academy of Sciences, 1970).

48 *Ibid.,* p. 44.

49 *Ibid.,* pp. 43–44.

50 I am referring here not to what Lamarck's successors found in his writings, but to what he himself wanted to suggest.

51 Louis Althusser, *Philosophie et philosophie spontanée des savants* (Paris: Maspero, 1967). SPS is the "spontaneous philosophy of the scientist," as distinct from COW, "conception of the world."

52 Canguilhem, *Idéologie et rationalité,* p. 9.

53 *Ibid.*, pp. 27–28.

54 This would be true if every instance of scientists working under the influence of Marxist theses resembled the Lysenko case. That is the question.

55 L. Althusser, *Philosophie spontanée*, pp. 99–100.

56 *Ibid.*, p. 100.

57 *Ibid.*, pp. 109–110 in particular.

58 *Ibid.*, pp. 108–109.

Chapter 4

1 Darwin uses the term in a sense that is still not very scientific.

2 I am borrowing the apt expression used by C. Limoges, "Lamarck et son milieu," *Science* 199 (1978), 1427–1428.

3 Lamarck, *Discours de l'an VIII, de l'an X, de l'an XI;* see Giard, ed., *Les Discours d'ouverture* (Paris, 1907).

4 *Discours de l'an VIII* in *Textes choisis* (Flammarion), p. 335.

5 Lamarck, *Philosophie zoologique*, p. xiv.

6 *Ibid.*, pp. 199–200.

7 *Ibid.*, part 1.

8 *Ibid.*, chapter 2.

9 *Ibid.*, chapter 3.

10 See chapter 5.

11 See the bibliography.

12 Lamarck, *Philosophie zoologique*, pp. xxxiff.

13 Francis Darwin, ed., *The Life and Letters of Charles Darwin* 2 volumes (New York: Appleton and Co., 1888), I, p. 384. The letter is dated 11 January 1844.

14 It is interesting that Cuvier himself defined the animal as a function of will.

15 *Avertissement* of the *Philosophie zoologique*, pp. xviff.

16 H. Graham Cannon, *Lamarck and Modern Genetics* (Manchester: Manchester University Press, 1959).

17 See the *Histoire naturelle* (1815 ed.), introduction, part 3, pp. 181–182.

18 Eloge of Cuvier (1835 ed.), p. xix.

19 H. G. Cannon cites the two texts juxtaposed, *Lamarck*, pp. 134–135. The Lamarck text is taken from the *Philosophie zoologique* (1873 ed.) I, p. 248.

20 Lamarck, *Philosophie zoologique* (1907 ed.), p. 200.

21 *Ibid.*, pp. 200–201.

22 *Ibid.*, p. 191.

23 *Ibid.*, p. 227.

24 *Idem.*

25 *Ibid.*, p. 229.

26 Genesis 30: 37–42.

27 Charles Darwin, *The Variation of Animals and Plants under Domestication* (London: J. Murray, 1868).

28 August Weismann, "On the supposed botanical proof of the transmis-

sion of acquired characters" (1888), in *Essays upon Heredity and Kindred Biological Problems,* transl. E. S. Schönland (Oxford: Clarendon Press, 1889), p. 387.

29 A. Giard, "Le principe de Lamarck et l'hérédité des modifications somatiques," *Controverses transformistes,* p. 135.

30 *Ibid.,* p. 138.

31 *Ibid.,* pp. 137–138; see also Weismann, "The supposed transmission of multilations" in *Essays upon Heredity,* p. 424.

32 See the bibliography.

33 Lamarck, *Discours,* in Giard, ed. (Paris, 1907).

34 Thus Grassé has edited the *Origin of Species.*

35 Wintrebert, *Le Vivant créateur de son évolution* (Paris: Masson, 1962), p. 15.

36 *Idem.*

37 *Ibid.,* pp. 16–17.

38 *Ibid.,* pp. 31ff.

39 *Ibid.,* p. 23.

40 P. Grassé, *L'Evolution du vivant* (Paris: Albin Michel, 1973), p. 24.

41 *Ibid.,* p. 10.

42 For example, the three volumes published by the University of Chicago.

43 See *La Recherche,* 89 (May 1978).

44 From the Department of Anthropology of the University of Geneva (kindly communicated by Albert Jacquard).

45 P. L. and J. S. Medawar, *The Life Science* (New York: Harper and Row, 1977), p. 95.

46 Jacques Monod, *Chance and Necessity,* transl. A. Wainhouse (New York: Vintage Books, 1972), pp. 109–110.

47 See Grassé, *Evolution du vivant,* pp. 364ff.

48 *Idem.*

49 *Ibid.,* p. 367.

50 *Ibid.,* p. 367, note 1.

51 H. G. Cannon, *Lamarck.* He evidently could not have known about them.

52 "As concrete as a stone or a pencil."

53 That is, an abstract factor that introduces an explanatory model.

54 C. H. Waddington, "Evolutionary adaptation," *The Evolution of Life,* I, p. 383.

55 This study is entitled "Le Généticien en proie à sa PSS," and appeared in *La Pensée,* 201.

56 See Jean Piaget, *Biology and Knowledge: An Essay on the Relations between Organic Regulations and Cognitive Processes,* transl. B. Walsh (Chicago: The University of Chicago Press, 1971), chapter 3, section 8. Piaget is mistaken when he speaks of "germen" and "soma" with regard to Lamarck; see pp. 121ff. and 134ff.

57 Dominique Lecourt, *Lyssenko. Histoire réelle d'une science prolétarienne* (Paris: Maspero, 1976).

58 *Ibid.*, pp. 126ff., especially p. 127, note 24, and p. 129.

59 G. Canguilhem, *Idéologie et rationalité,* p. 38.

60 *Ibid.*, p. 45.

Chapter 5

1 Camille Limoges, *La Sélection naturelle* (Paris: Presses Universitaires de France, 1970), chapter 2, p. 41.

2 C. Gide and C. Rist, *Histoire des doctrines économiques depuis les physiocrates jusqu'à nos jours* (Paris: Sirey, 1944), p. 90.

3 Lucien Cuénot, *L'Evolution biologique* (Paris: Masson, 1951), pp. 311ff.

4 Limoges, *Sélection,* p. 40. My italics.

5 See chapter 2.

6 Yvette Conry, *L'Introduction du darwinisme en France au XIXe siècle,* (Paris: Vrin, 1974), p. 306.

7 See A. Giard, *Controverses,* p. 138, and Lamarck, *Philosophie zoologique* (1907 ed.), p. 199.

8 Henri Bergson, *Creative Evolution,* transl. A. Mitchell (Westport, CT: Greenwood Press, 1975), p. 20.

9 The term "adaptation" is used here with the meaning given to it by L. Cuénot, and is retrospectively applied to a text that does not contain it.

10 Lamarck rarely uses the term *milieu.* When he does, its meaning is not that given to it by Darwin; it is essentially physicochemical, and has little to do with biology.

11 *Philosophie zoologique,* p. 188.

12 *Ibid.*, I, p. 222.

13 *Ibid.*, (1809 ed.) II, pp. 2ff.

14 *Ibid.*, II, p. 3.

15 *Idem.*

16 This tension is called "orgasm" (*orgasme*) and is a producer of irritability.

17 Lamarck, *Philosophie zoologique* (1809 ed.), II, p. 24.

18 Richard Burkhardt points out that Cuvier speaks of life as early as 1800. We have already mentioned Oken and Treviranus.

19 Quoted by B. Mantoy in his *Jean-Baptiste de Lamarck* (Paris: Seghers, 1968), p. 87.

20 *Idem.*

21 Richard Burkhardt, *The Spirit of System* (Cambridge, MA: Harvard University Press, 1977), pp. 58ff.

22 Lamarck, *Recherches sur l'organisation des corps vivants,* quoted by Wintrebert, *Le Vivant créateur de son évolution* (Paris: Masson, 1962), p. 29.

23 *Idem.*

24 *Ibid.*, p. 22.

25 For this set of questions, see C. Limoges, *Sélection naturelle,* pp. 35–

43. The quotation by Rostand is on p. 41.

26 In particular, Jean Rostand, quoted by C. Limoges, *Sélection naturelle,* p. 41, note 2.

27 Limoges, *Ibid.,* p. 39.

28 See Linnaeus, *Systema naturae* (1735).

29 Francis Darwin, ed., *The Life and Letters of Charles Darwin* 2 volumes (New York: Appleton and Co., 1888), I, p. 384.

30 The translator was probably R. Jameson, a professor of natural history; see H. G. Cannon, *Lamarck,* p. 15.

31 C. Limoges, *Sélection naturelle,* p. 41, note 2.

32 *Ibid.,* p. 42.

33 Letter to Hooker (11 January 1844), in F. Darwin, ed., *Life and Letters,* I, p. 384.

34 Charles Darwin, *On the Origin of Species.* [*Translator's note:* The original edition of Barthélémy-Madaule's book relies on French translations of Darwin. I have supplied the English text on which these translations are based, except in those cases where the book does not give enough information to identify the English text.]

35 Letter to Hooker (11 January 1844), in F. Darwin, ed., *Life and Letters,* I, p. 384.

36 *Ibid.,* I, p. 67.

37 Namely, that species descend from other species.

38 Charles Darwin, *On the Origin of Species* (London: J. Murray, 1859; Cambridge, MA: Harvard University Press, 1964), p. 3.

39 John Ray, *The Wisdom of God Manifested in the Works of Creation* (1692).

40 W. Paley, *Natural Theology* (1802).

41 See C. Limoges, *Sélection naturelle,* especially pp. 42–44; and R. Burkhardt, *Spirit of System,* pp. 72ff.

42 *Ibid.,* p. 46.

43 *Ibid.,* pp. 46–47.

44 Darwin, *The Descent of Man* (New York: Appleton and Co., 1888), p. 28.

45 Darwin, *Origin,* p. 8.

46 Darwin, *The Variation of Animals and Plants under Domestication,* 2 volumes (New York: Appleton and Co., 1892), p. 410.

47 *Ibid.,* p. 416.

48 Darwin, *Origin,* p. 60.

49 *Ibid.,* p. 81.

50 See for example the *Origin*, chapter 3.

51 *Origin* (Barbier translation), p. 232 [*translator's note:* translated from the French].

52 Y. Conry, *Darwinisme,* p. 193 *et passim.*

53 *Ibid.,* p. 39, note 40.

54 *Ibid.,* p. 191.

55 See chapter 4.

56 Y. Conry, *Darwinisme,* p. 306, note 78.

57 F. Jacob, *Logic of Life,* adapted from p. 149.

58 Y. Conry, *Darwinisme,* p. 425.

59 *Ibid.,* p. 193.

Chapter 6

1 R. Burkhardt, *The Spirit of System* (Cambridge, MA: Harvard University Press, 1977).

2 Goethe, *Conversations of Goethe with Eckermann and Soret* (London: George Bell and Sons, 1875), p. 480, hereafter Goethe, *Conversations.*

3 Arthur Schopenhauer, *On the Fourfold Root of the Principle of Sufficient Reason and on the Will in Nature,* transl. K. Hillebrand (London, 1888).

4 In this regard, see G. Canguilhem, *Études d'histoire et de philosophie des sciences* (Paris: Vrin, 1970), pp. 60ff.

5 See Y. Conry, *L'Introduction du darwinisme en France au XIXe siècle,* (Paris: Vrin, 1974), part 1, section 2, chapter 1.

6 In this regard, see Conry, *Darwinisme,* book 1, chapter 2.

7 *Ibid.,* p. 33.

8 *Ibid.,* p. 31.

9 *Ibid.,* p. 40.

10 Albert Gaudry was a well-known paleontologist who joined a teleological framework with admiration for Darwin.

11 Quoted by Y. Conry, *Darwinisme,* p. 238.

12 A. Giard, *Controverses transformistes* (Paris: Naud, 1903).

13 Lamarck, *Philosophie zoologique,* preface, p. 1.

14 E. Haeckel, *History of Creation* (New York, 1883), I, p. 111.

15 *Ibid.,* I, pp. 113–114.

16 See Goethe's text on nature at the beginning of the *History of Creation* (idealist texts used to good advantage by the materialist!).

17 See Haeckel, *History of Creation,* lesson 2.

18 His *Force and Matter* dates from 1885.

19 E. Haeckel, *The Proofs of Transformism* (an answer to Virchow).

20 G. Crespy, *La Pensée théologique de Teilhard de Chardin* (Neuchatel: Delachaux et Niestlé, 1965), p. 11.

21 Drummond, *Essai d'un naturaliste transformiste sur quelques questions actuelles* (1887), *Evolution et Liberté* (1885), and *Le Transformisme et le Récit biblique de la création* (1885).

22 See G. Crespy, *Teilhard,* pp. 18ff. and Y. Conry, *Darwinisme,* pp. 247–254.

23 Crespy, *ibid.,* p. 19.

24 See Y. Conry, *Darwinisme,* p. 234.

25 *Idem.*

26 Comte Maximilien Begouën, *La Création évolutive* (Toulouse: Privat,

1879), and *La Vibration vitale* (Tours: Rouille, 1885); Comte Begouën, *Quelques Souvenirs sur le mouvement des idées transformistes dans les milieux catholiques* (Paris: Blond et Gay, 1944).

27 Comte M. Begouën, *La Création évolutive,* p. 5.

28 *Ibid.,* p. 14.

29 *Ibid.,* p. 27.

30 *Idem.*

31 *Ibid.,* p. 43.

32 Begouën, *La Vibration vitale,* for example, pp. 84, 113.

33 Begouën, *Quelques Souvenirs,* pp. 20–21.

34 *Ibid.,* p. 25.

35 Begouën, *La Création évolutive,* p. 41.

36 *Ibid.,* pp. 41–42.

37 Y. Delage, *L'Hérédité et les grands problèmes de la biologie générale* (Paris: Schleicher frères, 1903), pp. 217ff. and 386ff.

38 Lucien Cuénot, *L'Adaptation,* (Paris: Doin, 1925).

39 H. Bergson, *Creative Evolution,* transl. A. Mitchell (Westport, CT: Greenwood Press, 1975), p. 86.

40 *Idem.*

41 *Idem.*

42 *Ibid.,* p. 271.

43 H. Bergson, *Les Deux Sources de la morale et de la religion* (Paris: Alcan, 1932).

44 Bergson, *Mélanges,* pp. 1162–1163.

45 A. Jacquard, *La Recherche,* September 1976.

46 *Idem.*

47 A. Jacquard, *Éloge de la difference: la génétique et les hommes* (Paris: Seuil, 1978).

48 Jacques Monod, *Chance and Necessity* (New York, 1972), p. 24.

49 Monod, *Leçon inaugurale,* p. 9.

50 Madeleine Barthélemy-Madaule, *L'Idéologie du hasard et de la nécessité* (Paris: Éditions du Seuil, 1972).

51 See the opening quotation of Jacques Monod's *Chance and Necessity.*

Conclusion

1 C. Limoges, "Lamarck et son milieu," *Science* 199 (March 1978), 1427.

2 See the famous example of the *rananculus,* which changes as a function of its aquatic or nonaquatic habitat.

3 G. Canguilhem, *Idéologie et rationalité* (Paris, 1977), p. 109.

4 *Ibid.,* p. 110.

5 *Ibid.,* p. 109.

Bibliography

1 Partial bibliography of the works of Lamarck according to Marcel Landrieu (1908)

1778 *Flore française* ou description succincte de toutes les plantes qui croissent naturellement en France, disposée selon une nouvelle méthode d'analyse et à laquelle on a joint la citation de leurs vertus les moins équivoques en médecine, et de leur utilité dans les arts. Paris, Imprimerie royale, 1778, in-8°, 3 volumes.

1783–1789 *Encyclopédie méthodique (Dictionnaire de botanique),* Paris, Panckouke, 1783–1817, in-4°, 8 volumes and 5 supplements.

1794 *Recherches sur les causes des principaux faits physiques,* et particulièrement sur celles de la combustion, de l'élévation de l'eau dans l'état de vapeurs; de la chaleur produite par le frottement des corps solides entre eux; de la chaleur qui se rend sensible dans les décompositions subites, dans les effervescences et dans le corps de beaucoup d'animaux pendant la durée de leur vie; de la causticité, de la saveur et de l'odeur de certains composés et de tous les minéraux; enfin, de l'entretien de la vie des êtres organiques, de leur accroissement, de leur vigueur, de leur dépérissement et de leur mort. Paris, Maradan, an II (1794), in-8°, with one plate, I(XVI + 315 pp.), II (412 pp.).

1798 "De l'influence de la lune sur l'atmosphère terrestre", *Journ. de phys.*, XLVI, 1798, 428–435; *Gilbert Annal.*, VI, 1800, 204–223; *Tilloch. Philos. Mag.* I, 1798, 305–306; *Bull. soc. philom.*, II, Paris, 1797, 116–118; *Nicholson's Journ.*, III, 1800, 488–490.

1800–1810 *Annuaires météorologiques,* Paris, 1800–1810, 11 volumes.

1801 *Système des animaux sans vertèbres,* ou Tableau général des classes, des ordres et des genres de ces animaux, présentant leurs caractères essentiels et leur distribution, d'après les considérations de leurs rapports naturels et de leur organisation, et suivant l'arrangement établi dans les galeries du Museum d'histoire naturelle, parmi leurs dépouilles conservées; précédé du Discours d'ouverture du cours de zoologie donné dans le Museum national d'histoire naturelle, l'an VIII de la République, le 21 floréal, Paris, Déterville, an IX (1801), in-8°, 452 pp. [A reproduction of the original edition was published in 1969.]

1802–1806 *Mémoires sur les fossiles des environs de Paris,* comprenant la détermination des espèces qui appartiennent aux animaux marins sans vertèbres et dont la plupart sont figurés dans la collection des vélins du Museum. Mollusques testacés dont on trouve les dépouilles fossiles dans les environs de Paris.

1803 *Histoire naturelle des végétaux* (with the collaboration of Mirbel), Paris, Déterville, an XI (1803), in-18°, 15 volumes.

1806 *Discours d'ouverture du cours des animaux sans vertèbres,* prononcé dans le Museum d'histoire naturelle en mars 1806, in-8°, 108 pp. (author's name not given). For a new edition see *Bull. Scient. France et Belgique,* XL, 1907, 105–157.

1809 *Philosophie zoologique,* ou exposition des considérations relatives à l'histoire naturelle des animaux, à la diversité de leur organisation et des facultés qu'ils en obtiennent, aux causes physiques qui maintiennent en eux la vie et donnent lieu aux mouvements qu'ils exécutent; enfin, à celles qui produisent les unes les sentiments et les autres l'intelligence de ceux qui en sont doués. Paris, Dentu, 1809, I (XXV + 428 pp.), II (475 pp.) in-8°. [An edition of this work was published in 2 volumes in 1969.]

1815–1822 *Histoire naturelle des animaux sans vertèbres,* présentant les caractères généraux et particuliers de ces animaux, leur distribution, leurs classes, leurs familles, leurs genres, et la citation des principales espèces qui s'y rapportent; précédée d'une introduction offrant la détermination des caractéres essentiels de l'Animal, sa distinction du végétal et des autres corps naturels; enfin l'exposition des principes fondamentaux de la zoologie. Paris, Déterville, 1815–1822, in-8°, 7 volumes. [An edition of this work was published in 7 volumes in 1969.]

1820 *Système analytique des connaissances positives de l'homme* restreintes à celles qui proviennent directement ou indirectement de l'observation. Paris, Belin, 1820, in-8°, 352 pp.

2 Selected list of sources for Lamarck's biography

a. Official documents

Institut royal de France—Académie royale des sciences. Funérailles de M. le chevalier de Lamarck (12 mm., Paris, 1830). Discours de Latreille, p. 1 : discours de E. Geoffroy Saint-Hilaire (au nom du Museum), p. 5. Museum d'histoire naturelle au jardin du Roi. Funérailles de M. le chevalier de Lamarck. Discours prononcé par M. le chevalier Geoffroy Saint-Hilaire le 20 décembre 1829 (8 pp.).
Cuvier G., Éloge historique de M. de Lamarck, read before the Académie des sciences (by Baron Sylvestre) le 26 novembre 1832. (Mém. Acad. sc., XIII, 1835) and *Recueil des éloges historiques* (Paris, Didot, 1861; new ed. III, 189–210).

Partially reproduced by Bourguignat, with a bibliography and 3 portraits, *Revue bibliographique de la Société malacologique de France* (Paris, 1886, 61–95) and by Guillemin, *Archives de botanique* (I, 86–95, 1833), with a commentary.

b. Various studies and articles on Lamarck and his doctrine

Bonnet Ed., *"L'Herbier de Lamarck,* son histoire, ses vicissitudes, son état actuel," *Journ. de botan.,* XVI, 1902.

Brooks and Cunningham, "Lamarck and Lyell: Lyell and Lamarckism" *Nat. Science,* VIII.

Brooks W. K., *The Foundations of Zoology,* 1899.

Bureau, *L'Herbier de Lamarck,* C. R. Acad. sc., 17 January 1887.

Clos M., *Lamarck botaniste :* sa contribution à la méthode dite naturelle et à la troisiéme édition de sa flore, Mém. Acad. Sc. belles-Lettres et arts de Toulouse (9) VIII, 1896.

Cope E. D., *The Primary Factors of organic evolution,* 1896.

Duval M., *Le Transformiste français Lamarck,* Société anthropologique, 7th conférence, 1889.

Duval, M., *Le Darwinisme,* Imprimerie Hennuyer, 1889.

Geoffroy Saint-Hilaire, I., *Histoire naturelle générale des règnes organiques,* Paris, Masson, 1854–1862.

Giard A., *Controverses transformistes,* Paris, Naud, 1903.

Giard A., *Les Idées de Lamarck sur la métamorphose* (reply to Ch. Pérez), C. R. soc. biol., 1903, p. 8.

Giard A., "L'évolution dans les sciences biologiques", *Bull. scient. France et Belgique,* XLI, 1907.

Giard A., Avant-propos à la réimpression des Discours d'ouverture des cours de zoologie de J. B. Lamarck, *Bull. scient. France et Belgique,* XL, 1907, p. 443.

Giard A., Présentation des Discours d'ouverture de Lamarck à la Société de biologie, *C. R. soc. biol.,* LXXII, 1906, p. 319.

Geikie A., "La géologie au début du XIXe siècle : Lamarck et Playfair", *Revue scient.,* 16 June 1906.

Hamy E. T., *Les Derniers Jours du jardin du Roi et la fondation du Museum d'histoire naturelle,* Imprimerie nationale, 1893.

Hamy E. T., *Les Débuts de Lamarck,* Paris, Burdin, 1907.

Hermanville, "Notice biographique sur Lamarck, sa vie et ses œuvres", *Bull. soc. Acad. de l'Oise,* XVII, 1898.

Lacaze-Duthiers H. de, "De Lamarck", *Revue cours scient.,* 16, 1866.

Landrieu M., *"Lamarck et ses précurseurs", Revue école anthrop.,* XVI, 1906, p. 152.

Landrieu M., "De Lamarck à Darwin", *Revue des idées,* 15 July 1906.

Landrieu M., "Les origines et la jeunesse de Lamarck", *Revue Scient.,* November 1907.

Lanessan J. L. de, *Le Transformisme, évolution de la matière et des êtres vivants,* Paris, Doin, 1883.

Lanessan J. L. de, *Transformisme et Créationnisme,* Paris, Alcan, 1914.

Le Dantec F., *Lamarckiens et Darwiniens,* Paris, Alcan, 1899.

Le Dantec F., *La philosophie zoologique de Lamarck* (introduction aux Limites du connaissable), Paris, Alcan, 1903.

Locard A., "Une visite aux collections malacologiques de Lamarck", *Échange,* 44, 1888.

Lyell, *Principes de géologie* (French translation by Ginestou, 1873, of *Principles of Geology*.
Martins Ch., "Un naturaliste philosophe : Lamarck, sa vie, son œuvre", *Revue des Deux Mondes,* 1873, with biographical introduction to the new ed. of *Philosophie zoologique,* 1873.
Packard A. S., *Lamarck, the Founder of Evolution. His Life and Work,* Lingmons, Green, 1901.
Perrier E., *La Philosophie zoologique avant Darwin,* Imprimerie nationale, 1884.
Perrier E., *Lamarck et le Transformisme actuel,* Imprimerie nationale, 1893.
Quatrefages A. de, *Charles Darwin et ses précurseurs français,* Paris, Baillière, 1870.
Royer C., "Lamarck", *Revue phil. positiv.,* III, 1868; IV, 1869.

3 Principal works on Lamarck subsequent to Landrieu's bibliography

Althusser L., *Philosophie et philosophie spontanée des savants* (1967), Paris, Maspero, 1973.
Aron J.-P., "Les circonstances et le plan de la nature chez Lamarck", *Essais d'epistémologie biologique,* 1969.
Aristote, *Traité de l'âme,* Ed. Budé.
Begouën M. (comte), *La Création évolutive,* Toulouse, Privat, 1879.
Begouën M. (comte), *La Vibration vitale,* Tours, Rouillé, 1885.
Begouën M. (comte), *Quelques Souvenirs sur le mouvement des idées transformistes,* Paris, Blond et Gay, 1944.
Bergson H., *L'Évolution créatrice,* Paris, Alcan, 1941.
Bergson H., "La philosophie" (1915), in *Écrits et Paroles,* Paris, PUF, 1959, t. II.
Blanc L., *Histoire de la Révolution française,* Paris, Librairie du Progrès, 1866.
Bonnet Ch., *La Palingenésie philosophique,* in *Œuvres complètes,* Neuchâtel, Fauche, 1779–1783, t. XV-XVI.
Bonnet Ch., *La Contemplation de la nature,* in *Œuvres complètes,* Neuchâtel, Fauche, 1779–1783, t. VIII-IX.
Bourdier F., "Lamarck et Geoffroy Saint-Hilaire face au problème de l'évolution biologique", *Revue d'histoire des sciences,* 25, 1972.
Bourdier F., *L'Homme selon Lamarck,* colloque international Lamarck, 1971.
Boyer F., "Le Museum d'histoire naturelle à Paris et l'Europe des sciences sous la Convention", *Revue d'histoire des sciences,* 26, 1973.
Burkhardt R., "Lamarck, evolution and the politics of science", *Journ. of the History of Biology,* 1970.
Burkhardt R., *The Spirit of System: Lamarck and Evolutionary Biology,* Harvard University Press, 1977.
Burkhardt R., "La théorie de Lamarck sur la conduite des animaux et des hommes", in *Lamarck et son temps, Lamarck et notre temps,* Paris, Vrin, 1981.
Canguilhem G., *La Connaissance de la vie,* Paris, Vrin, 1969.
Canguilhem G., *Études d'histoire et de philosophie des sciences,* Paris, Vrin, 1970.
Canguilhem G., *Idéologie et Rationalité,* Paris, Vrin, 1977.
Cannon H., *Lamarck and Modern Genetics,* Westport, CT, Greenwood Press, 1959.
Comte A., *Cours de philosophie positive.*

Conry Y., *L'Introduction du darwinisme en France au XIXe siècle,* Paris, Vrin, 1974.
Crespy G., *La Pensée théologique de Teilhard de Chardin,* Paris, Éd. universitaires, 1961.
Cuénot L., *L'Adaptation,* Paris, Doin, 1925.
Cuénot L., *L'Évolution biologique,* Paris, Masson, 1951.
Cuvier G., *Éloge à l'Académie, 1830.*
Darwin C., *L'Origine des espèces,* présentation de P. Grassé, Paris, Marabout-Université, 1973 (French translation of *On the Origin of Species*).
Darwin C., *Théorie de l'évolution,* selections by Y. Conry, Paris, PUF, 1969.
Daudin H., *Cuvier et Lamarck : les classes zoologiques et l'idée de série animale,* Paris, 1926.
Dumont L., *Haeckel et la Théorie de l'évolution en Allemagne,* Paris, Baillière, 1873.
Delage Y., *L'Hérédité et les Grands Problèmes de la biologie générale,* Paris, Schleicher frères, 1903.
Geoffroy Saint-Hilaire E., *Discours,* Paris, Firmin-Didot, 1829.
Gide C. et Rist C., *Histoire des doctrines économiques,* Paris, Sirey, 1944.
Gillispie Ch., *Lamarck and Darwin in the History of Science,* Baltimore, 1959.
Gillispie Ch., "The Formation of Lamarck's Evolutionary Theory,"*Archives internationales d'histoire des sciences,* 1956.
Gould S., "Lamarck et la paléontologie américaine", in *Lamarck et son temps, Lamarck et notre temps,* Paris, Vrin, 1981.
Grassé P., "Lamarck, biologie," *Revue scientifique,* 1944.
Grassé P., *Lamarck et son temps. L'évolution,* Paris, Masson, 1957.
Grassé P., *L'Évolution du vivant,* Paris, Albin Michel, 1973.
Goethe, *Conversations de Goethe, recueillies par Eckermann,* Paris, Charpentier, 1865.
Haeckel E., *Histoire de la Création,* Paris, Schleicher frères, 1879.
Haeckel E., *Les Preuves du transformisme,* Paris, Baillière, 1879.
Jacob F., *La Logique du vivant,* Paris, Gallimard, 1970.
Jacquard A., *Éloge de la différence* (la génétique et les hommes), Paris, Éd. du Seuil, 1978.
Hilary Rose, Steven Rose et al., *L'Idéologie de/dans la science,* Paris, Éd. du Seuil, 1977.
Landrieu M., *Lamarck, le fondateur du transformisme,* Mémoires de la Société zoologique de France, Paris, 1908.
Lecourt D., *Lyssenko. Histoire réelle d'une science prolétarienne,* Paris, Maspero, 1976.
Le Dantec, *La Philosophie zoologique de Lamarck,* Paris, Alcan, 1904.
Limoges C., *La Sélection naturelle,* Paris, PUF, 1970.
Limoges C., "Lamarck et son milieu", *Science,* 30 March 1978.
Mantoy B., *Jean-Baptiste de Lamarck,* Paris, Seghers, 1968.
Monod J., *Le Hasard et la Nécessité,* Paris, Éd. du Seuil, 1970.
Paley W., *Natural Theology,* Rivington, 1821.
Perrier E., *Lamarck,* Paris, Payot, 1925.
Piaget J., *Biologie et Connaissance,* Paris, Gallimard, 1970.
Prenant M., *Biologie et Marxisme,* Paris, Hier et aujourd'hui, 1948.
Russo F., *Pour une bibliothèque scientifique,* Paris, Ed. du Seuil, 1972.
Schelling, *Idées pour une philosophie de la nature* (1797), et *Introduction à la*

première esquisse d'un système de la philosophie de la nature (1799), in *Essais,* Paris, Aubier, 1946.
Schiller, *L'Échelle des êtres et la Série chez Lamarck,* colloque international Lamarck, 1971.
Schopenhauer A., *La Volonté dans la nature,* Paris, PUF, 1969.
Weismann, *L'Hérédité,* Paris, Schleicher frères, 1883.
Wintrebert, *Le Vivant créateur de son évolution,* Paris, Masson, 1962.

Index